My Philosophy

*Representing my Views on the Many
Functions of the Ether of Space*

OLIVER LODGE

CAMBRIDGE
UNIVERSITY PRESS

CAMBRIDGE UNIVERSITY PRESS

Cambridge, New York, Melbourne, Madrid, Cape Town,
Singapore, São Paolo, Delhi, Mexico City

Published in the United States of America by Cambridge University Press, New York

www.cambridge.org
Information on this title: www.cambridge.org/9781108052672

© in this compilation Cambridge University Press 2012

This edition first published 1933
This digitally printed version 2012

ISBN 978-1-108-05267-2 Paperback

CAMBRIDGE LIBRARY COLLECTION

Books of enduring scholarly value

Physical Sciences

From ancient times, humans have tried to understand the workings of the world around them. The roots of modern physical science go back to the very earliest mechanical devices such as levers and rollers, the mixing of paints and dyes, and the importance of the heavenly bodies in early religious observance and navigation. The physical sciences as we know them today began to emerge as independent academic subjects during the early modern period, in the work of Newton and other 'natural philosophers', and numerous sub-disciplines developed during the centuries that followed. This part of the Cambridge Library Collection is devoted to landmark publications in this area which will be of interest to historians of science concerned with individual scientists, particular discoveries, and advances in scientific method, or with the establishment and development of scientific institutions around the world.

My Philosophy

In his study of optics, Newton postulated that light, like sound, must be carried through a medium, and that this medium must exist even in a vacuum. By the late nineteenth century, this theoretical substance was known as the luminiferous ether. But the ether theory faced several problems. If the earth moved through ether, there would be ether wind, and light travelling against the flow would move more slowly than light travelling with it. That was soon disproven. Nor could the ether be stationary: by 1905, Einstein's work on relativity had disproven absolute motion. In this fascinating advocacy of ether, first published in 1933, Sir Oliver Lodge (1851–1940) fiercely defends ether against the new physics, arguing for solid models over mathematical abstractions, and urging new ether experiments. With in-depth references to Einstein, Jeans and Eddington, this book is still relevant to students in the history of science.

Cambridge University Press has long been a pioneer in the reissuing of out-of-print titles from its own backlist, producing digital reprints of books that are still sought after by scholars and students but could not be reprinted economically using traditional technology. The Cambridge Library Collection extends this activity to a wider range of books which are still of importance to researchers and professionals, either for the source material they contain, or as landmarks in the history of their academic discipline.

Drawing from the world-renowned collections in the Cambridge University Library and other partner libraries, and guided by the advice of experts in each subject area, Cambridge University Press is using state-of-the-art scanning machines in its own Printing House to capture the content of each book selected for inclusion. The files are processed to give a consistently clear, crisp image, and the books finished to the high quality standard for which the Press is recognised around the world. The latest print-on-demand technology ensures that the books will remain available indefinitely, and that orders for single or multiple copies can quickly be supplied.

The Cambridge Library Collection brings back to life books of enduring scholarly value (including out-of-copyright works originally issued by other publishers) across a wide range of disciplines in the humanities and social sciences and in science and technology.

MY PHILOSOPHY

MY PHILOSOPHY

REPRESENTING MY VIEWS ON THE
MANY FUNCTIONS OF

THE ETHER OF SPACE

by

SIR OLIVER LODGE

" We will grieve not, rather find
Strength in what remains behind;
In the primal sympathy
Which having been must ever be;
In the soothing thoughts that spring
Out of human suffering;
In the faith that looks through death,
In years that bring the philosophic mind."

WORDSWORTH: *Intimations of Immortality.*

LONDON
ERNEST BENN LIMITED
1933

FOREWORD

The Ether of Space has been my life study, and I have constantly urged its claims to attention. I have lived through the time of Lord Kelvin with his mechanical models of an ether, down to a day when the universe by some physicists seems resolved into mathematics, and the idea of an ether is by them considered super-fluous, if not contemptible. I always meant some day to write a scientific treatise about the Ether of Space; but when in my old age I came to write this book, I found that the Ether pervaded all my ideas, both of this world and the next. I could no longer keep my treatise within the proposed scientific confines; it escaped in every direction, and now I find has grown into a comprehensive statement of my philosophy.

OLIVER LODGE.

9th February, 1933.

CONTENTS

PART ONE

AN ELEMENTARY SURVEY OF PHYSICAL EXISTENCE

7

CONTENTS

CONTENTS

CHAPTER VIII

MACHINERY OF GUIDANCE

Animated matter is controlled by the agency of life, and an organism has the power of renewing itself, and directing the energy which it finds in space without interfering with its amount. Some physical agent is needed to explain the control, and it is urged that the explanation is not complete without taking the ether into account. All our interpretation of existence is involved. . . . .

PART TWO

EVIDENCE FOR AND CONTROVERSIES CON- CERNING THE ETHER

CHAPTER IX

MATTER, ENERGY, AND THE ETHER

Fundamental experience gradually growing gives us all three things, the ether being that part of space in which the energy resides.—Energy is transmitted from one piece of matter to another without being perceived in transit.—Electric and magnetic fields require the same medium as light. Locomotion is characteristic of matter, but its statement and measurement are a demonstration of etherial activity. . . . .

CHAPTER X

THE ETHER AND THE FORMS OF ENERGY

We have no dynamics of the ether at present: our senses have only told us about matter.—Force is a reality between particles of matter.—Speed and fatigue give us our idea of time.—Potential energy resides in the ether, and is just as real as the other kind called kinetic. . . . . .

CONTENTS

CHAPTER XI

FARADAY'S CONCEPTION OF THE ETHER

CHAPTER XII

MODERN GIBES AT THE ETHER

CHAPTER XIII

THE PHYSICAL ASPECT OF THE UNIVERSE AN ALTERNATIVE SCHEME TO THAT OF SIR JAMES JEANS

CONTENTS

CONTENTS

CONTENTS

13

CONTENTS

CONTENTS

15

PART ONE
AN ELEMENTARY SURVEY OF PHYSICAL EXISTENCE

B

CHAPTER I

THE CONSTITUTION OF THINGS AROUND US. A SURVEY OF EXISTENCE

"Yet I doubt not thro' the ages one increasing purpose runs,
And the thoughts of men are widen'd with the process of the suns."
TENNYSON—*Locksley Hall.*

This Universe in which we find ourselves is an infinite reservoir of Possibilities. We have scratched its surface on the material side—the side which appeals to us through our senses—and have penetrated some of its secrets, but the amount which we do not know, even on this side, is portentous: science is a comparatively recent development of humanity. We now find that the material universe is built of two apparently fundamental entities—or at most three, not yet resolved into one—the positive and negative units of electric charge, and the photon, or unit of etheric radiation.

How all the complexity of visible existence can be thus composed, is amazing; but there seems no doubt that so it is. The two electric units, the proton and the electron, in regular number and order of movement, compose the atoms of matter; and of them every material substance is built up. The multiplicity of elements, discovered and studied by the great Chemists of the past and present century, are law-abiding aggregates of those two fundamental electric units; and by their number and arrangement in the atom the properties of each kind of atom are on the way to be explained. The material universe bids fair to be reduced to a glorified sort of mechanism controlled by electric forces, just as the heavenly bodies have long been known to be controlled, and their motions regulated, by the equally or still more mysterious force of gravitation.

There is indeed a third thing, in which these electric units exist and of which they are probably composed, a uniting and welding entity which is responsible for electric and gravitational and all other forces; but this third entity makes no appeal to our senses, and has till comparatively recently been ignored; for, though undoubtedly physical in its properties, it is not a part of the material universe in the strict sense. All we have effectively known of it is radiation; and as to the exact composition of that there is still some uncertainty. We do suspect, however, that it is through radiation that atoms act on each other, and that it is through some kind of radiation that the otherwise chaotic assemblage of electrons is welded into a cosmos, on a large scale, and into crystals and other solids and liquids, on a small scale. Radiation is purely an affair of the ether of space, and we know that the ether is the vehicle of electric and magnetic and gravitational and cohesive forces, as well as of radiation. It is through radiation that all our acquaintance with distant bodies has come; and but for our sensitive receiving instrument the eye—which enables us readily to appreciate certain kinds of radiation—we should be in a still denser state of ignorance.

Existences which do not affect the eye or any of our senses seem to us to be non-existent, and inferences about them are sometimes thought to be mere superstition. Real superstition, or gratuitous baseless fancy, is indeed a deadly and disreputable weakness; but the reasoning faculty which enables us to make legitimate inferences, and the intuitions and inspirations with which some of the race are favoured, are in no sense despicable,—these indeed are among the higher facul-

A SURVEY OF EXISTENCE

ties of man. They are the faculties which more than anything else distinguish him from the rest of the animal kingdom. On his bodily and sensual side he is akin to the animals, but on his mental and spiritual side he soars above them till he is near akin to the Divine.

By the Divine I mean the superhuman controlling and guiding Power which regulates the cosmos, as we in our insignificant way regulate the conduct of ordinary affairs. Only gradually and very imperfectly can we form any conception of what that Power is like or what its attributes are.

Instinctively we know that we are not the highest of all the Beings in existence. The revelations of science,— the fact that all that we are now aware of has been going on from time immemorial, long before humanity appeared on the scene, and the certainty that we have had no hand in bringing all this beauty and adaptation and marvellous structure into existence,—suffice to show, or enable us to infer, that there must be grades of existence higher as well as lower than man; and it is reasonable to suppose that such grades of existence extend upward at least as far as they extend downward; they can hardly have a limit short of infinity. The Infinite Being we call God, and we seek after him if haply we may find some token of his Presence, some notion of his Nature, some indication of his Purpose and Aim.

In this search we may make blunders and go astray; but we are encouraged to do our best, and are conscious of help from time to time vouchsafed by the Higher Powers with whose will and intention we reverently try to co-operate.

CHAPTER II

THE PROGRESS OF PHYSICAL SCIENCE

"There is nothing more obstructive to the advance of knowledge than a certain unformulated dogma implicitly accepted by many men of science, namely, the dogma that what we cannot fully understand cannot happen. We cannot too strongly insist that the bounds of the possible do not coincide with and are not set by the limits of our present powers of comprehension."

PROFESSOR McDOUGALL.

The history of the material universe has been worked out by the great modern astronomers through immense epochs of time. The origin of satellites from planets, and of planets from a sun, have been traced back to physical causes which can be followed. Suns in enormous number exist in space, and the origin of these has also been traced to an immense whirling mass of gas called a spiral nebula. These nebulæ we see in various stages of evolution, and they are reckoned to be the birthplace of stars. The nebulæ themselves are great aggregations of gas brought together by gravitation from a more primitive condition. The most primitive condition which has as yet been envisaged by science is a number of atoms or particles distributed uniformly throughout space. This uniform distribution could not be stable, and could not continue for ever. A homogeneous distribution of matter throughout space, if it ever existed, would tend to coagulate into a number of separate masses, and ultimately would condense into a multitude of nebulæ, with great spaces between them.

22

EVOLUTION

This represents the beginning of a course of evolution formulated by Herbert Spencer as a change from a homogeneous uniformity into a heterogeneous crowd. According to his formula "In evolution matter passes from an indefinite incoherent homogeneity to a definite coherent heterogeneity." Or in simpler words, Evolution is from the simple to the complex, from the unorganised to the organised, or from a featureless continuity to an assemblage of atoms, and so on, to definite structures obeying certain rules and subject to law and order. Like the change from invisible vapour to a cloud, and then on to the raindrops which are formed of its material and fall towards the earth where they contribute to the growth of vegetation. The great masses of gas formed in space have only begun the course of evolution, but the law of their development after the stage of formation is the subject matter of the whole of cosmological, it may even be said of physical, science. One of these masses has given rise to the whole of our stellar system: and one of the stars so formed has given birth to a system of planets, on one of which we live, and are beginning to trace its long and complicated evolutionary history.

But now, going back to the beginning, which is supposed to be the ether of space with a number of particles of matter embedded in it, distributed uniformly: that is no ultimate beginning. It contains a vast store of energy, which must be conserved in all the subsequent processes, and will serve to maintain its activity through the whole length of its career. It may some day be possible to postulate the form of that energy as a minute or fine-grained vortical circulation of the continuous ether throughout its whole vast extent—an internal rotatory motion characterised by the enormous velocity c. And it may be possible to make reasonable, what is already a surmise, that that fine-grained circulation may under certain conditions generate the electrons and protons of which the atoms of matter are composed, so that matter should, as it were, crystallise out of an unmodified spatial ether, the original seat of all the energy in the universe. According to this idea matter becomes the palpable part of the ether—the only portion of it which affects our organs of sense, and therefore the only portion which is incontrovertibly *known* to us. It is the portion of the ether which we handle and deal with every day. But even so, we have not got at an origin, and I see no way of getting an origin by attending to the material universe alone. If the ether, and that which is born from it, is to be the vehicle of spiritual development, we must assume that it has been inhabited or utilised by a psychical or spiritual entity all along, and that the interactions of mind with matter, which display themselves so frequently to our senses, had begun and were already existing in that dim and distant past. So that a cosmogonist of the present day, like Sir James Jeans, writing as a physical astronomer or astronomical physicist, can say, If you want to go further into the original cause, so as to account for the circulation of the ether itself, you can think of the Finger of God stirring it up. We can trace the physical operations back and back as far as we can, but not without limit. Sooner or later we arrive at something which is not physical, which has more analogy with our minds than with our bodies, and which we sometimes call idealistic and sometimes spiritual. It is this portion to which I wish to call attention.

In connexion with the attempt to imagine a turbulent motion in the ether of space as the origin or precursor of the whole material universe, it may be urged that the use of such an expression as "the Finger of God" or "the Creative Word" or "the Divine Brooding" is an abandonment of all attempt at a scientific explanation,

EVOLUTION

and a falling back on a mode of expression utterly outside and beyond the range of physical science. I entirely agree, but my thesis is that sooner or later this appeal to a higher and more idealistic aspect of the universe is inevitable. Science is utterly incompetent to explain the existence of the world as we know it now. Existence itself is a problem beyond its scope. If we can trace back the present condition of things to a dim and distant past, in which they were not so complicated, and can realise that they have developed out of a much simpler and more homogeneous condition, we have exhausted the powers of physical science, in the quest for origins, and can proceed no further.

We are as yet a long way from having done as much as that; but the efforts of the great astronomers of the present day, combined with the researches of the mathematical physicists, tend to bring an interpretation of that kind somewhere within the bounds of at least speculative possibility. The origin of a solar system, the birth of the stars, and even the previous condition of the great spiral nebulæ, seem to be coming within our reach. There is some close connexion between radiation and corpuscular particles, which we have begun to recognise, and which suggests that the primary material of physics is etheric wave motion, or some kind of cellular vortex structure in a fundamental continuum, occupying and perhaps constituting all space; time also being involved in the postulated angular momentum of each unit cell and in its power of developing into something more interesting and more complex in the course of evolution.[1]

Manifestly here is no ultimate origin. The existence of an etheric whirl has to be accounted for just as strin-

[1] The suggestion is that corresponding to each cell is the quantum unit of angular momentum $h\,2\pi$; while the volume of a cell is somehow related to h^3

gently as the existence of the complex universe to-day. There is no need to suppose that in that original turmoil the whole of its future development was secretly locked up and latent.

Evolution

One idea of evolution is that it is a mere unfolding of what is already essentially there, so that all the happenings in the world were latent in the fiery cloud or nebula to which it owed its origin. On this view, nothing is added from the outside beyond certain particles of matter, which can be accreted from other bodies co-existent with itself. So that if you take the whole material aggregate into consideration, you have the possibility of everything that the organism can subsequently exhibit. Some have extended this to the animate creation, and to the works of man, so that the whole achievement of humanity has to be considered latent in the stormy whirl of a primeval cloud. I contend that this doctrine is entirely untrue. A child may develop into a great man, but the great man was not latent in the child. A multitude of other influences have to enter into his composition, not only material but spiritual, and not till they are all assembled does the great personality appear.

Whatever truth there may be in pre-existence (and I by no means deny it), it is absurd to fix upon one particular portion, and claim that to be the whole. The nebulous or gaseous mass once supposed to have been broken off from the sun by tidal action, is only one ingredient that has gone to the making of all the multifarious phenomena on the planet earth. It contained maybe all the atoms of matter; but it contained little else. There was much energy in a potential form, which has since become actual, or rather, manifest; and there must have been in existence a multitude of spiritual influences,

EVOLUTION

which ultimately became incarnate or incorporated with the matter, but which previously gave no sign, and were not taken into consideration. Creatures are apt to begin in a formless mass, and end with an elaborate structure. Only the material part of the organism existed in the formless mass: the guiding and directiong principle entered into it from outside, and was not apparent until it did. Evolution ought to take the whole into account, and not be regarded as a mere unfolding of unseen possibilities, unless we take the unseen region also into the evolutionary balance-sheet. It is impossible that all the complexity of the world should have been latent in the matter at the start. Interaction and enrichment has continually occurred all along. Energy has conspicuously come into it from outside through the etheric activity of radiation. Whether the spiritual influence has entered in in analogous etheric fashion, I cannot say; but that it has entered in is certain.

All that we need admit about the course of evolution from a primeval nebula is that there was the potentiality of material structures and of all their manifold energies. There is no need to adopt the words of Omar Khayyám in exaggerated fashion, as if they represented not only the doctrine of determinism but also the idea that nothing fresh was ever introduced, so that every subsequent development must be supposed implicity contained in every previous condition:—

"With earth's first clay they did the last man knead,
   And then of the Last Harvest sow'd the Seed:
Yea, the first Morning of Creation wrote
   What the Last Dawn of Reckoning shall read."

We need not believe that. We can keep the way open for fresh powers to enter from the psychical side; for a kind of emergent evolution where the emergent novelty does not emerge solely from the previous physical state. It is conceivable that even from the Highest point of view the universe may be a great Adventure. This view is inspiring, for it suggests that to the extent of our ability we can help in the Project, and thus in some manner share the Divine Responsibility.

CHAPTER III

DESIGN AND PURPOSE IN THE UNIVERSE

"Man is a spiritual being; the proper work of his mind is to interpret the world according to his highest nature, to conquer the material aspects of the world so as to bring them into subjection to the spirit."

ROBERT BRIDGES.

The same step outside the material universe that was needed to explain its existence may be made equally well when we try to account for its animation and subsequent elaboration, in accordance with a principle of Design to a purposed end. If we have to postulate a spiritual world at all, we may as well utilise it throughout; not appealing to it unnecessarily, seeking always a proximate explanation; holding on to physics as far as we possibly can, but being ready to abandon it whenever its methods are seen to be entirely incompetent. The abandonment, or rather the transcendence, of pure physics has already been found necessary in dealing with the behaviour of animated creatures. Live things, it is now generally admitted, are actuated by something more than the physical and chemical reactions of their material organisms. A scientific explorer from another planet, examining the earth at some future stage in its history, could not account for the remains or ruins of the roads, the bridges, the houses, the churches, as the result of physics and chemistry alone. He would have to postulate the activity of a race that had designed and planned these things, and constructed them for some specific

29

purpose. In other words, he would have to admit an idealistic interpretation of the earth as now we know it. So even on this little planet experience shows that a mental and spiritual world has been active; every piece of machinery, no matter how automatic, shouts that it has been designed for some planned and foreseen end. And that which is conspicuous on a small scale may be extended without breach of continuity to the greatest things of which we have cognizance.

It is familiar that a psychic element enters into our everyday experience; and often we are aware that there must have been some plan or design or purpose associated with each simple observation, although we may be unable to specify what it is. Thus, for instance, in taking a country walk recently with my daughter, we noticed in the valley a long row of willows that had just been pollarded, but among them we saw two trees standing in the row which had been uncut. The natural ejaculation was, What are those two left for? Why are they treated differently from the others? There must have been some purpose. That is the point. The feeling is inevitable that there must have been some purpose. As far as the mechanism was concerned, the cutting or non-cutting of the trees was quite straightforward, but there was a psychic element of some simple and obvious kind, which nevertheless was not apparent. This is an absurdly simple illustration of what is familiar enough, that we constantly have to make appeal to the spiritual world, or at least to some psychic entity of a quite ordinary kind, for the complete explanation of any simple experience. I do not know why those trees were left, and it is no concern of mine, though doubtless if it were worth while I could find out somebody and ask: which again is a psychic operation, and indeed is the method we naturally employ for finding our way across an unknown country. We are continually immersed in a psychic or spiritual world, though our touch with it is so commonplace that we ignore the mystery attaching to it. Our very speech, our writing, is a psychical as well as a physical phenomenon. And only very exceptionally does the occurrence of these two aspects of existence attract our notice, or seem to need further exploration. All the common objects around us in a room are full of human design and purpose. They are signs of human ingenuity as well as of an application of physical energy obtained from food, and utilised in a way studied by physiologists as a matter of nerves and muscles.

The explorer from another world could probably infer the laws of energy on which the things have been constructed, could reckon the amount of work done in their erection, and might even reinvent the machinery that had been employed. He would have no difficulty with the laws of energy, but he would see that something more was necessary, and that an element of design and purpose, that is of some mental or spiritual activity, was necessary, as well. So it is that we infer from their fossil remains the existence of live creatures who disported themselves on the earth in inaccessibly distant periods of its history. So also we infer from the carvings and decorations of prehistoric works of art the existence of an intelligent race with some approach to culture. A work of beauty and design hands down to posterity a world of meaning, and can be far more instructive than are the physical laws of energy involved in its construction. I claim that the material universe with its variously designed atoms, and the way they have been used in the construction of all the objects, mineral, vegetable, and animal, that we see around us, is a sign also of gigantic Design and Purpose, and is a glorious Work of Art.

Accordingly, I say that when we come to philosophise on existence at any stage, we not only may, but we must, transcend the limitations of physical science, even in its broadest range, and admit the working and operation of a superhuman guiding and directing Power. Sooner or later, though we try to avoid such a step, as we trace the history backwards, we all find that we are bound to make it. We cannot understand the existence either of ourselves or of an external world unless we postulate some kind of creation. Creation involves design and purpose and mental activity, and necessarily involves a creator of some kind. We appeal to exemplifications, on a minute scale, in poems and music and works of art. They are human creations, and we know something of their creators. But whether we know them or not, whether they are lost in the mists of antiquity or whether the varnish on their works is not yet dry, no one doubts that a creator they must have had, and that to seek to explain their inner meaning and purpose solely in terms of the behaviour and arrangement of the atoms would be absurd. Yes and if the thing we find in a Cretan palace or an Egyptian tomb should happen to be a domestic implement or a machine constructed for some specific purpose, we can take an interest in analysing that purpose and reconstructing the civilisation of which it forms a part. It is a sign or symbol or manifestation of some mental state, which in some cases may be higher or more advanced than we should have thought likely in those prehistoric times.

In dealing with the universe as a whole we have no prehistoric qualms to contend with, no hesitation about attributing Intelligence to the operations of a distant Mind or Logos. "In the beginning was the Word": the Mind responsible is still active to-day, and we have no reason to suppose that it has changed in the least. The material universe has evolved, and has rendered possible a fresh influx of spiritual reality as it attained greater complexity, but the Creator may be the same yesterday

DESIGN AND PURPOSE

to-day and for ever. His Design and Purpose in bringing the Universe into existence may not be apparent to us; or we may form some hazy conception of it. That is a relative and subjective matter, not of much consequence except to ourselves. But surely we may have faith that there is a Design and Purpose running through it all, and that the ultimate outcome of the present cosmos, and all its manifold puzzles, will be something grander, more magnificent and more satisfying than anything we unaided can hope to conceive. That has been the faith of poets, and that I hope may be the faith of the states-men who enter into the turmoil and carry on the little business of humanity from day to day. By that faith they may be strengthened in their task, and feel that to the extent of their opportunity they are helping to work out one corner of the Majestic Scheme.

If we try to limit ourselves to material considerations alone, we may get depressed, downhearted, and pessi-mistic. We do not and cannot see the ultimate outcome, and may be afflicted with doubts as to whether there is to be any permanent outcome. But if we are con-strained to admit the activity of a spiritual world, let us be consistent in that admission. Let us trace and utilise its activity throughout, and make use of all the knowledge and help vouchsafed. It is the effort of religion to utilise this help and knowledge; and, in spite of many mistakes and blunders, the religious feeling in its essence, and apart from forms and cere-monies, is as alive and energetic as ever it was. Philosophy justifies it, and science contributes an element towards its justification. This is the meaning of the great Cathedrals and of the forms of Religion. They teach that the spiritual world is the great reality: all else, however beautiful and interesting, is temporary and evanescent. The universe is ruled by Mind, and whether it be the Mind of a Mathematician or of an

c   33

Artist or of a Poet, or of all of them, and more, it is the one Reality which gives meaning to existence, enriches our daily task, encourages our hope, energises us with faith wherever knowledge fails, and illuminates the whole universe with Immortal Love.

CHAPTER IV

RELIGION AND SCIENCE

"Science cannot impugn the affirmation of the supreme value of the spiritual; but it may deny, and through the mouths of many of its leaders it has denied, that the spiritual is of any effect in the life of man. It cannot deny that man recognizes and acclaims truth, goodness, and beauty in all their forms; but it has denied that man's aspiration to conserve and to create these values is of any efficacy. And this is the most fundamental part of the attack of Science upon Religion. For, if this denial is well founded, Religion is revealed as wholly illusory."

PROFESSOR McDOUGALL,
Author of *Mind and Body*.

It is often said that there can be no conflict between the two great departments of human interest, called Religion and Science, because they deal with different themes in totally different ways, and therefore never overlap, so that there is no possibility of a fight,—the kind of thing that used to be said of nations before 1914. But this is an exaggeration; no human being can always be satisfied with any one department of knowledge; there are times when he must seek a more comprehensive view. The alternative is to keep your thoughts from straying out of the region to which you are attending, and concentrate on one side only at a time. This has been the practice of many good people, and it is certainly one way of avoiding conflict. But it is not a philosophic or permanently satisfying way. Between the two great regions there is a frontier, and there squabbles and conflicts may occur, unless there be some

35

way of reconciling the frontier inhabitants of the two regions. Our senses tell us that there is a material world full of things which we can investigate: our instincts and intuitions tell us that there is a mental and spiritual world where also we can feel at home. But the object of this book is to show that there is also an etheric or metetherial world, which spreads over both the other regions, and may be the means of reconciling them or of enabling us to attend to either or both without dislocation or shock. We find that the material world by itself is inert and inactive and also full of discontinuities, and that its operations cannot be understood unless we take into account a continuum, which may be treated as an abstraction and not studied in detail, or else as a fundamental substance with physical properties, to whose interaction with matter all our familiar activities are due, but which in itself is not well adapted to physical investigation. My hypothesis is that this same almost unexplored physical substance operates also in the region of life and mind, and ultimately will be found to be the physical vehicle utilised in the spiritual region, so as to constitute the mechanism whereby spirit and matter interact, and that it will ultimately form a bond of union between the two domains experimentally known to humanity.

The methods used in the two departments are certainly very different, especially when exemplified by the attitudes of their extreme supporters on either side. The religious man cultivates his feelings of faith and awe-struck wonder by acts of contemplation and worship. He traces the Finger of God in action everywhere. His instinct is to appeal for spiritual assistance in any trouble. He is not concerned with the method by which operations are conducted, and he sometimes regards the effort made to explain the process by which ordinary results are achieved as too mechanical and atheistic a method to be at all satisfactory or even permissible. The more extreme members of this school of controversy are so imbued with the notion of Divine activity that they need no other explanation, and consider the formulation of a material explanation as perverse and irreligious. They trace the direct action of God, not only in the operations of nature, but in human writings, and consider that in those writings they have a system of infallible Authority, which they have only to study in order to arrive at all truth.

There is another equally extreme set among the devotees of science, who, by close application to the detailed phenomena which appeal to the senses, are so satisfied with what they have learned about the complexity of the working and the completeness of the explanation, that they are ready to exclude every other mode of expression, and consider all attempts in that other direction the outcome of baseless superstition. They realise the complexity of the mechanism; they perceive that effects are always produced by changes and adaptations in the material organisms. They never ask for the object or purpose involved, or for the why and wherefore of the devices they study, but are satisfied with what they call the " how " of the actions: until they gradually get to believe that a narration of the physical processes involved is not only all that they can attend to, but all that exists; so that they feel furious and contemptuous with those of the opposite side who recognise and seek to ascertain Divine Intention and Purpose.

Between these extreme groups there is an intermediate set of people, who have in some respects a more difficult, and in others an easier and more peaceful, task. They recognise a truth on both sides. They know that they are part of a material frame of things, yet they never deny that they are also in touch with a mental or

spiritual world: they feel that in their innermost personality they are already denizens of that world. Some
of them can communicate with it; and they are all ready
to ask for help and guidance in any emergency. At the
same time they know perfectly well that every action is
produced in accordance with physical laws, and that
without proper mechanism, no activity such as appeals to
the senses is possible. They know that their knowledge
is extremely imperfect; and they are not tempted to deny
either the mechanical explanations of the one group, or
the spiritual aspirations and ideals of the other. They
deny nothing, but seek to learn from both sides whatever
contributions to truth they can receive.

To this intermediate group, in its various grades,
probably the majority of the human race belong; and
the higher more intelligent grade of this group probably
contains most of the philosophers who try to take a comprehensive view of the universe as a whole, including
both the material and the spiritual side. They do not try
to keep them separate, but admit their interaction. In
the phenomena of animated matter they perceive the
interaction at work in its simplest and most accessible
form. They welcome the teachings of the materialistic
biologists on their positive side, and rejoice in their
discoveries. The discovered laws of heredity and the
powers of adaptation in living creatures give them
information, and do not perturb them. They realise
that a process of evolution has actually gone on, and that
it is the means whereby the richness and variety of the
material forms have been produced. When in a scientific mood, they are as loth as are the others to appeal to
the Finger of God or to any spiritual agency, as part of
the mechanism to be appealed to in order to get an
understanding of the process. They know that such
appeal is illegitimate in science, and is equivalent to
admitting defeat; yet they do not deny that such opera-

tions are continually occurring, and they believe that for a comprehensive understanding of everything, including Design and Purpose, such appeal must be made. Their doctrine is that in the last resort, for an ultimate explanation even of the simplest thing, the Divine Will must be evoked. Consequently they can reconcile the two modes of thought by saying that every natural process is conducted in accordance with a spiritual intention and a far-seeing design, but yet that the operations are conducted according to a system of mechanical law and order, that can gradually be disentangled, formulated, and more or less understood. They do not look for sudden irruptions of divine activity, or for any breach of laws. They realise that Divinely designed mechanism would be perfect in its working, and require no tampering with, no suddenly executed repairs, nor any other introduced improvements, such as are familiar in all but the most perfect kind of human machinery. They believe that a physical universe has been contrived, so as to be perfect in its working, except in so far as it is interfered with and modified, or it may be spoilt, by the free will agency of other spirits, who, being associated with material organisms for a time, have a certain power of interfering with it, and bringing about results which, though foreseen, cannot be considered part of the Divine Intention.

To this group the doctrine of an etheric intermediary or continuum, fulfilling the purposes of interaction and serving as the means whereby results are achieved (however gradual or law-abiding those results may be)— this doctrine I say should be a help, and enable the operations both of the material and the spiritual world to be understood more fully and more comprehensively than without such intermediate agency.

I have argued that, according to our experience in physics, the ether is continually in operation, and that all our activities are due to the interchange of energy between ether and matter, whether it be brought about by automatic mechanism or as the result of our own purpose and design. And as a speculation I surmise that the idealistic and essentially real permanent part of ourselves is in continual touch with this semi-etheric mechanism, if so it can be called, and is the means whereby even the simplest spontaneous actions are accomplished. We ourselves have no power of causing any change or any movement in the material world except through the agency of an organism. And by study we find that the atoms of matter are operated on by the ether, and by nothing else. All physical fields of force exist in the ether, and must be effective in every change produced in inert particles of matter, whether they be parts of an inorganic substance or parts of an animated body.

We thus find that all activity is to be sought for in space. Activity is not to be explained in terms of matter alone, nor is its range of action to be limited to the lumps of matter which we see around us. Astronomers tell us that space is far more extensive than the discontinuous material bodies which occupy an infinitesimal portion of its vast extent. And I surmise that this continuum is the home or habitat of all that we call life and mind, and constitutes its body or means of manifestation. The ether is the only means whereby spirit can act on matter, the only means whereby its activities can be brought within the range of our human senses. It is indeed the only means whereby matter can be acted on at all. The argument to be developed is that matter *per se* is quite inert and initiates no sort of change.

CHAPTER V

THE ORGANISM AND THE CONTROL

"Religion asserts rather that in such experiences man makes contact with an aspect of the Universe that is real and supremely important, an aspect which takes precedence of the physical realm. Furthermore, religion assumes that he not only makes contact with this realm but also shares in it, partakes of it, is influenced by it, and in return can contribute something, however little, to it."

PROFESSOR MᶜDOUGALL.

There is so much that is still mysterious about this interaction, and so about what is called the normal behaviour of any ordinary person; but a few individuals show signs of an unusual kind of mental activity; they do not behave quite normally. Their organism may be affected by something akin to disease, or they may be influenced in other more wholesome but more puzzling ways. Most people seem to have a uniform commonplace personality, which continues uninterruptedly to animate their organism from the cradle to the grave; all have a period of unconsciousness in sleep; and some show a kind of dislocation of personality, it may be a duplex variety, whereby, although the ordinary course of life is managed by one, every now and then another personality takes control; and in those cases the occasional interloper *may* have greater powers of intelligence, and may operate the mechanism to a better and more valuable result than the ordinary normal control can. Of this enhancement of personality there are many grades. During the access of what in the higher forms is called genius, the ordinary commonplace personality is semi-oblivious, or it may be completely oblivious, of the common affairs of life: while the outcome of the exalted mood is a poem or drama or symphony, such as in his ordinary state the individual could not have composed. The process is called "inspiration" in the higher stages, and sometimes "possession" in the lower. Both the conscious and the subconscious parts of the mind appear to be affected by the process; it is of all grades, and various degrees of value. To the higher grades of this phenomenon we owe most of the supreme works of art which humanity treasures. The lower grades we find exemplified in the peculiar condition which we have only recently begun to recognise and study under the name "mediumship." There are still lower grades, which call for the attention of psychiatrists, and which may be summarised under the contemptuous term "lunacy." But whatever their grade, or whatever their value may be, they are all examples of the interaction of a spiritual world, or rather of a psychical influence of some kind, not commonly familiar in the ordinary experience of commonplace humanity.

A few scientific persons have begun to attend to these exceptional manifestations, and try to make out the laws under which they act, to dissect out the coherence and uniformity subsisting among the incoherencies and exceptional character of these occurrences. I shall not enlarge further upon that branch of the subject, now, but shall refer to the great work of F. W. H. Myers on *Human Personality*, called by its full title *Human Personality and its Survival of Bodily Death*. For Myers, like others, was convinced by a study of the phenomena that man as a whole was much greater than his bodily organism, that his character and behaviour, his memory and affection, could not be explained in terms of that organism alone: rather that the organism was constantly animated by something from the immaterial world, something which does not appeal to our senses, and in its essence is essentially a spirit associated for a time with matter, in order to acquire earthly experience and achieve earthly results. Furthermore he felt assured that his existence continued uninterruptedly, together with the character, memory, and affection which he had acquired here, and which could last long after he had discarded the bodily instrument which he had put together out of material particles and used for purposes of demonstration while still in the flesh. That was his view, the view that he gradually found to be the only one consistent with the facts. That also is my view, and the view of a growing number of others. I rather hope that the group who call themselves Rationalists, and who pride themselves on being loyal to fact, may make a serious study of these facts before long. It is a fertile field that has been left too long unexplored.

But doubtless there are those who disagree; and so long as they admit the customary interaction of an unseen and unsensed idealistic activity, which controls, dominates, and actuates all the particles of animated matter which we find on this planet, that is all I demand for the present purpose. I am convinced that the phenomena of nature hang together as a consistent whole, and that what we have ordinarily observed as an interaction between the psychical and the physical, is a deep-seated phenomenon which runs through all existence, and which contains the clue to the working of the whole universe. We invariably find that mental or psychical activity is demonstrated by some physical activity, the result of physical energy directed along some path indicative of design and purpose. Hence it is reasonable to suppose that this dependence on the physical is a necessary condition of psychical activity. And accordingly we always tend to look for a physical agent which can be made use of by life and mind for the purpose of bringing about changes in the material universe.

The instinct of the materialists has been thereby largely justified. They always sought for a physical agent to account for every action. Their only weakness was to be satisfied with that, to require nothing more, and even to ignore the action if no physical agent could be found. My view is that a physical agent will always be necessary for a complete explanation, that every phenomenon is psycho-physical, but that the physical agent involved may be inconspicuous and need some drawing out from its hiding place.

CHAPTER VI

THE PROPERTY OF INERTIA

Inertia must be regarded as a fundamental property of matter; and it is important though not easy to get our ideas clear about it. The property is so fundamental that it is rather difficult to define it. It is commonly thought of as something akin to laziness; and this is correct, provided we extend the term "laziness" to every kind of change, and apply it equally to a disinclination to change of motion as well as to a disinclination to change from rest. It represents, in truth, an inability rather than a disinclination. Matter has no power of changing its state, whether of rest or motion, nor indeed doing anything else to effect a change in its behaviour. If a top is spinning, it must continue to spin at the same rate for ever, unless it is acted upon by some force. A body possessing this property if in motion must continue to move with the same speed unaltered, unless affected by some external agent. Inertia has therefore been called persistence. But persistence suggests the idea of activity or active exertion, which is the very thing that it is intended to deny to matter. It only persists in motion because it has no power of doing anything: it has not even the power of stopping itself. This is so different from our experience of the behaviour of any matter known to us that it is not easy to grasp the full significance of it. To show how easily the fact may be mis-stated, a quotation from the

45

great work in physics called *Thomson and Tait's Natural Philosophy* may be cited:—

"Matter has an innate power of resisting external influences, so that every body, as far as it can, remains at rest or moves uniformly in a straight line."

This suggests that the inertia of matter enables it to resist forces which are not applied to it. Whereas I want to emphasise that it has no power at all. Clerk Maxwell, in his review of Thomson and Tait's great book, seizes upon this statement, and gibes at it thus:—

"Is it a fact that 'matter' has any power, either innate or acquired, of resisting external influences? Does not every force which acts on a body always produce exactly that change in the motion of the body by which its value, as a force, is reckoned? Is a cup of tea to be accused of having an innate power of resisting the sweetening influence of sugar, because it persistently refuses to turn sweet unless the sugar is actually put into it?"

And he goes on to say that we must get "rid of this Manichæan doctrine of the innate depravity of matter, whereby it is disabled from yielding to the influence of a moving force unless that force actually spends itself upon it,"—a statement with which the distinguished authors would of course agree. Their statement about inertia was erroneous. Helplessness or inability to change or initiate anything whatever is rather the significance that has to be attributed to inertia. Unless a mass is acted upon in some way by an external agent, it has no power of doing anything. It is entirely passive. Its reaction to a force is essential to the existence of a force upon a body, but it is completely passive: it is entirely different from the active resistance of an animal. Inertia means the complete absence of any activity, yet it would be impossible to exert any force

INERTIA

upon a body which possessed no inertia. A cart reacts on a horse with the same force as the horse acts on the cart. They are two bodies and one force acts on each. Reaction is not opposition, it is a necessity of action. If the cart did not react, the horse would be useless. Imagine a horse harnessed to a piece of cottonwool or fluff. Effort is impossible without reaction, the two are *always* equal. Inertia is the ingredient which confers momentum upon a body; it is a factor of momentum, the other factor being speed. It is this factor which enables motion to continue, and so confers kinetic energy upon a body. In order to trace a few elementary consequences of this fact, I will make extracts from a Friday evening discourse which I gave at the Royal Institution on the subject of inertia so long ago as February 1919.

We are each of us flying through space at nineteen miles a second, probably much more. Nothing is propelling us; we continue to move by our own inertia, simply because there is nothing to stop us. Motion is a fundamental property of matter. No piece of matter is at rest in the ether, the chances are infinite against any piece having the particular velocity zero; every bit is moving steadily at some given speed, unless acted on by unbalanced force. Then it is accelerated—changed either in speed or direction, or both.

As a matter of fact we, like other bodies on the earth, are acted on by two slight unbalanced forces—one which makes us revolve round the earth once a day, like a satellite, the other which makes us revolve round the sun once a year, like a planet or asteroid. Our annual revolution is not because we are attached to the earth; we are not attached, but revolve as independent bodies, and would revolve in just the same time and way if the earth were suddenly obliterated: only then we should find the diurnal revolution transmuted into a twenty-four hour rotation round our own centres of gravity, and the eccentricity of our annual orbit very slightly changed. In any case there is no propelling force, only a residual radial force producing curvature of path.

A railway train, or a ship moving steadily, is likewise subject to no resultant force. Propulsion and resistance balance. The whole power of an engine, after the start, is spent in overcoming

47

friction. The motion continues solely by inertia. Any steadily moving body is an example of the first law of motion. You need not try to think of a body under no force at all; you cannot think of such a body on the earth, but you can think of one under no resultant force, i.e. under balanced forces. Such a body moves by reason of its inertia alone. It is in equilibrium; it is not at rest.

But we have no sense of straightforward locomotion, and not the slightest clue to either the magnitude or direction of our motion through space. We can ascertain approximately how the sun is moving with reference to our system or cosmos of stars, but we do not know at what rate that system is itself moving. For all we know it may be moving very fast; hundreds of miles per second.

We have a sense of acceleration however; we experience it in a lift as it begins to descend; and if the sensation is repeated often enough, as on a rough sea, the result is unpleasant. We have also a sense of rotation; we can tell when our vehicle—say a Tube train—turns a corner in the dark. Most animals appear to have a sense of rotation, apparently located in the ear. But we have no sense of direct translation; and we have so far failed to devise any instrumental means for detecting our motion through the ether of space.

The failure is not for lack of trying. Many experiments have been tried, but there is always some compensating effect; so we get no answer to the question—at what rate and in what direction we are moving? The best known experiment is that of Michelson and Morley, the result of which seems to assert that the ether clings to the earth, or that the earth is not moving through any kind of substance. But Fitzeau's classical experiment showed that a transparent body carried with it none of the external ether of space; and experiments made by myself[1] at Liverpool in the 'nineties of last century show that a rapidly moving opaque body carries no external ether with it, that there is no perceptible viscous drag or cling between matter and ether, and accordingly demonstrates that stagnation or absence of relative ether drift past the earth is not a reasonable explanation of Michelson's negative result.

The two experiments together, in fact, ought to be taken as establishing the reality of the most interesting of all the compensating effects yet discovered, viz. the FitzGerald-Lorentz contraction of all matter in motion, which the electrical theory

---

[1] See Phil. Trans., vol. clxxxiv. (1913), pp. 727–804, and vol. clxxxix. (1897), pp. 149–166.

of cohesion renders so extremely probable. It only amounts to a 3-inch shrinkage in the whole diameter of the earth in the direction of motion; but it is enough. This slight contraction or change of shape in moving bodies I regard as the definite and interesting compensating effect in this case. Incidentally, moreover, it establishes the electrical, i.e. the chemical, nature of cohesion. For given that cohesion is a residual chemical affinity —due to the outstanding attraction of molecules composed of neutral groups of equal opposite electric charges, brought so near together that the attraction between molecules is no longer averaged to zero[1]—then, on orthodox Maxwellian electric theory, a diminution of this force due to lateral motion is inevitable. And the resulting lateral expansion or longitudinal contraction, or both, is of the right order of magnitude. So this acts as a previously quite unsuspected compensating effect, which exactly neutralises the drift effect otherwise to be anticipated. Thus, by superposition of two positive consequences of drift, the Michelson experiment, like every other yet made, declines to indicate that there is any drift at all.

Hence, after many such negative results, it seems to become hopeless to enquire experimentally as to our motion through ether. Unless indeed gravitation were exempt from the otherwise universal compensation. In that case the electrical theory of matter applied to the motion of planets might yield a residual result. But my enquiry into this problem has suggested that gravitation too is in the conspiracy,[2] and in that case, there is some ground for the contention of the extreme Relativists, not only that we do not know our motion—with which everyone agrees—but that we never shall know it: and, in fact, that motion of matter through ether is a phrase without meaning.

I hope we shall not too readily shut the door on further attempts in this direction; and as a conservative physicist I may be allowed to lament the extraordinary complexity introduced into physics and into natural philosophy by the principle of relativity, as so remarkably and powerfully developed by the mathematical genius of Einstein, with complication even of our fundamental ideas of space and time.[3] The complications do not commend themselves to all of us, and I for one should be glad to return to the pristine simplicity of Newtonian dynamics, modified of course by the electrical theory of matter; admitting the

[1] See for instance my book on Electrons, chap. xvi.
[2] See *The Philosophical Magazine* for August, 1917, and February, 1918, pp. 145, 155 and 156.
[3] This was written before the verification of relativity theory, but I leave it as a confession of prejudice.

MY PHILOSOPHY.—PART I

FitzGerald-Lorentz contraction, and admitting also the variation of effective inertia with speed. These things do not destroy, but supplement, Newtonian dynamics. They generalise it in a legitimate and intelligible manner. Such complications as these are clearly in accordance with truth and are to be welcomed; but the complicated theory of gravitation created this century by Einstein, and developed by his successors, and the consequent overhauling of space and time relations, do not at present commend themselves to me, nor I think to others of what I suppose must be called the older school.

Meanwhile the full-blown theory has the courage of its conviction and has predicted a definite result, viz. the deflexion of a ray of light by the sun's limb, equal to 1·75 seconds of arc. The prediction is going to be tested during the solar eclipse of May 29 this year, between Brazil and the Gulf of Guinea. Let the issue be clearly understood. If a star-ray grazing the sun is deflected $\frac{3}{4}$ second it will mean only that light has weight, that the wave-front not only simulates the properties of matter by carrying momentum—as we know it does from the investigations of Nichols and Hull, Poynting and Barlow, and others—but that it is even subject to gravity. For this would be the angle between the asymptotes of a cometary orbit when the comet is moving with the speed of light and passing close to the sun.[1] But the principle of relativity—through the refractive or converging influence of a strong divergent gravitational field—demands a greater deflexion than this, more than twice as great. So there are three alternate deflexions before us, to be settled by observation:—

1·75 sec.; 0·75 sec.; and zero.

Let us hope that the result of this or of some other eclipse-opportunity may be definite enough to discriminate clearly and quantitatively between these three alternative values; any one of which should be equally welcome to any lover of truth.

If the first answer is given decisively, it will be a conspicuous triumph for the theory of relativity, and will for a time be hailed as a death-blow to the ether. I claim beforehand that such a contention is illegitimate, that the reality of the ether of space depends on other things, and that the establishment of the principle of relativity leaves it as real as before; though truly it becomes even less accessible, less amenable to experiment, than we might have hoped. Nevertheless the ether is needed for any

[1] See, for instance, my paper in *The Philosophical Magazine* for August, 1917, page 93.

INERTIA

clear conception of potential energy, for any explanation of elasticity, for any physical idea of the forces which unite and hold together the discrete particles of matter whether by gravitation or cohesion or electric or magnetic attraction, as well as for any reasonable understanding of what is meant by the velocity of light. Let us try to realise the position beforehand; for we shall be handicapped in the progress of our knowledge of the relation between matter and ether until these fundamental things are settled, and until everyone agrees that the ether has a real existence. I want people generally to admit that the ether is itself stationary as regards locomotion, and that it is the seat of all potential energy; and further, at least as a surmise, that it is the medium out of which matter is probably made, and in which matter is perpetually moving by reason of its fundamental property called inertia—a property the full explanation of which must, I expect, ultimately be relegated to and considered as a property derived from the ether itself.

Is there anything else, besides matter, which possesses or seems to possess inertia! Faraday discovered that an electric current had a property which bore some analogy to inertia, a property clearly depending on its magnetic field. Every current, even a convection current, is necessarily surrounded by lines of magnetic force, and when the magnetic field is intense the current behaves as if it had considerable inertia. Faraday at first called the effect "the extra current." Maxwell called it "self-induction." The latter is the better name.

To show it I start a current in a circuit containing a stout ring of laterally subdivided iron round which the current-conveying wire is wound, and I put in circuit an instrument which only responds when the current has risen to nearly its full strength. A current usually rises, what is called, instantaneously, but here there is a very noticeable delay between pressing down the key and the response of the instrument. The lag shown is only a second or two, but with care I can adjust it till it is a quarter of a minute. Such delay or lag in establishing a current would be fatal to electric telegraphy. In practice the delay is reduced to a minimum, by using its early values, and the actual response is exceedingly quick. Still, the law of rise of current is quite definite, there is no exception, it is only a question of degree; and the law is the same as that appropriate to the pulling of a barge on a canal. A barge gets up speed slowly, at a rate depending on its mass or inertia, and it ultimately attains a steady speed when the resistance balances the pull.

That is exactly the case of a steady current obeying Ohm's law,

51

MY PHILOSOPHY.—PART I

the E.M.F. is balanced by the resistance, the propelling force is zero, and the current flows by what we may call its own inertia—its own momentum.

To stop the current you must either increase the resistance or suspend the propelling force. If you interpose an obstacle suddenly, the motion stops with violence—a collision in the case of a train or barge, a flash in the case of electric current. This is what Faraday called "the extra current at break," and if you are holding the wires in your hand when a current is suddenly broken in a circuit of large self-induction you may get a nasty shock.

If you could abolish electric resistance a current would go on for ever without propelling force.

An amazing experiment has been made by Kamerlingh Onnes at Leiden, who first cooled a metal ring down to within four degrees of absolute zero by means of liquid helium and then started a current through it by a momentary magnetic impulse. Instead of stopping in a minute fraction of a second, as usual, the current went on and on, not for seconds but for days. In four days it had fallen to half strength, and there were traces of it a week later. A most suggestive experiment as to the nature of metallic conduction, as well as a demonstration of the fly-wheel-like momentum of an electric current!

This electromagnetic analogue to mechanical momentum or inertia is explicable (or supposed to be explicable) in terms of the magnetic field surrounding the current, i.e. really (as I think) in terms of a property of the ether of space. It exactly simulates inertia; but is it an imitation or is it the same thing? Can it be said that an electric charge possesses inertia in its own right, and retains it always, as matter does, whether it be moving or whether it be stationary?

The question was brilliantly answered by your Professor of Natural Philosophy, Sir J. J. Thomson, so long ago as 1881. He calculated the inertia or quasi "mass" of an electric charge e, on a sphere of radius a, and showed that it was $m = 2 \mu e^2/3a$.

The μ need not be attended to now, though it is really the most important of all—being a great etherial constant of utterly unknown value[1]—but for our present purpose the μ merely signifies that the e must be measured in electromagnetic not electrostatic measure, when the formula is interpreted numerically with $\mu = 1$.

At the date 1881 this expression for true electric inertia,

---

[1] I have guessed that it is a density of 10^{11} grammes per c.c. $\div 4\pi$. See *The Ether of Space*, Appendix 2; also the *Phil. Mag.* for April, 1907.

INERTIA

though an interesting result, seemed too absurdly small to have any practical significance. Take a sphere like a football, 20 centimetres or 8 inches in diameter; charge it till it is ready to give more than an inch spark, say up to 60,000 volts; then calculate the inertia or equivalent mass corresponding to the charge. If I have done the arithmetic right it comes out one-third of a millioneth of a millioneth of a milligramme (3×10^{-16}) Absurdly small! Yes, but not zero. And whenever a quantity is not nothing, there is no telling what importance may not have to be attached to it sooner or later. Nothing real can be so small as to be really negligible in the long run as knowledge progresses. Something at present unforeseen may bring it into prominence. So it has turned out in this case. The infinitesimal result of nearly forty years ago to-day dominates the horizon. It was in some sort the dawn of a new era in physics.

Consider it further. Clearly the inertia depends not on the charge only, but on its concentration. The radius of the sphere occurs in the denominator of the expression. The same charge on a sphere 2 centimetres in diameter would have ten times the inertia; on a sphere as small as an atom the inertia would be a hundred million times bigger still. But then even that is small; moreover an atom could scarcely be expected to hold such a charge. Nevertheless, allowing only a reasonable potential, it might seem that atomic inertia could be sensibly increased by an electric charge. But no, even on a sphere as small as an atom the concentration turns out insufficient; the effect is still excessively minute. Yet as electric inertia at given potential depends on linear dimensions, while material inertia depends on those dimensions cubed, there must be a size when the two are equal, i.e. when one might account for the other.[1]

---

[1] Few things are more surprising than the extraordinarily large charge held by or constituting an electron in proportion to its size. The charge is so large that ordinary arguments about electricity as it exists on material spheres cannot be expected to apply. If they did, or in so far as they do, the potential of an electron would not be two volts but well over a million volts; and the density of the etherial substance of which it is presumably composed (if its electric inertia is to be derived in any simple ordinary way from its bulk), would have to be nothing like that of water, but of the order 10^{11}, or a billion times the density of water. A thousand tons, in fact, to the cubic millimetre.

We are here out of our depth among quantities on which a great deal of work has to be done to reduce them to order. Yet it must not be supposed that these figures are nonsensical. They require serious consideration; and that is all that can be said for them. I do not think there is any sense in talking about the potential of an indivisible unit of charge, but we can talk about the potential existing at the confines of an atom; and that is a reasonable magnitude, about 14 volts in the case of hydrogen, and not very different for other elements.

But on the other side of the subject everything points to the density of

MY PHILOSOPHY.—PART I

The moral of this elementary argument is that not for bodies of atomic size, but for something 100,000 times smaller in linear dimensions, is it possible to explain inertia electromagnetically. But, 40 or even 20 years ago, one would have said—there are no bodies of this size; nothing can be smaller than an atom! The strange thing is that, as nearly everyone knows now, bodies of this size have been discovered. They were isolated by Sir J. J. Thomson in 1897, having been gradually led up to by Crookes's and many other experiments on cathode rays; and they are shown to be an apparently invisible unit or atom of electricity whose inertia is wholly electric.

Furthermore it was found that the very same electrons can be split off or detached from any or every kind of atom, that there is only one kind of negative electron; and though at first there appeared to be many kinds of positively charged particles, the evidence is tending to the discovery of a single kind of positive electron likewise; so it is natural to suppose that electrons are an essential ingredient in matter. And since they possess inertia, even those which are clearly disembodied electric charges, it becomes possible to surmise that in some sense, or in a certain grouping, they constitute the atom, that they confer upon it the inertia with which we are familiar and that in fact electric inertia is the only inertia that exists.

Electric inertia began as the simulacrum of material inertia, it has shown itself the very same thing, and it seems likely to end by displacing every other kind of inertia altogether.

This is the electrical theory of matter.

―――――

Assuming this theory for the present as a working hypothesis, we may say that material inertia is explained electromagnetically, i.e. is explained in terms of the magnetic field which necessarily

―――――――――――――――――――――

ether being exceedingly high, though perhaps not so high as the above estimate. It must at least be greatly denser than platinum or lead, and probably immensely denser.

A difficulty is often felt as to how ordinary matter like a planet can move through such a medium without friction. Density, however, does not involve viscosity; the two are disconnected; and resistance to motion would be caused only by viscosity, of which the ether appears to have none. There are many ways, more or less satisfactory, of picturing the perfectly free motion of matter through an exceedingly substantial ether of space; there would be innumerable difficulties in supposing friction and consequent generation of heat. It is quite certain that whatever the ether does it does not dissipate energy. That imperfection belongs to the province of molecularly constituted matter alone.

54

INERTIA

surrounds and accompanies every charge in motion; since a charge in motion constitutes a current. For on this view a material body is but an aggregate of such charges grouped according to some definite pattern, positive and negative charges interlaced or somehow intertwined, and so far apart in proportion to their size that they do not interfere with each other or cancel each other, nor apparently overlap or encroach on each other's field, to any measurable extent. Is this possible? It is. For comparing the size of an electron with the size of an atom we perceive that they are relatively of the same order as the size of a planet and the size of a solar system. So it becomes possible to think of an atom as a sort of solar system, with a positive nucleus or sun surrounded by negative electrons revolving in regular orbits round it.

On this view, or indeed in any form of the electrical theory of matter, the atom of matter consists mainly of empty space; in other words, it is excessively porous; just as the solar system is mainly empty space and may be spoken of as excessively porous; the actual material lumps being almost infinitesimal in proportion to the total bulk. A rapid projectile or a ray of light passing through the solar system would be unlikely to hit anything, the chances would be strongly against a collision. So also, if a point be thrown through an atom, the chance of its hitting anything is about 1 in 10,000. It might pass through 10,000 atoms before striking. This experiment has been tried, by C. T. R. Wilson and others, and that is roughly speaking the result. Sooner or later a radium projectile meets with an obstacle and is stopped, but it traverses a good number of atoms on the average; it traverses quite a perceptible distance even in a dense solid, before it strikes a nucleus.

Matter accordingly seems to me—to us I may say, for in this most physicists are I think agreed—a gossamer or milky-way structure, an impalpable accident in the substantial ether. Here a speck and there a speck, but, for the great bulk of it, empty space!

"Impalpable" is not the right word, for matter is essentially palpable. It is because it appeals so directly to our senses that we attend to it so vividly. It forces itself on our attention, while the ether eludes us. And why? Clearly because our bodies are composed—our sense organs are composed—of this very matter. On the material side we are part of, and thoroughly at home in, the material universe. Whereas the ether is elusive; we know nothing of it directly; and though our eyes are instruments for receiving etherial tremors excited by agitated electrons, we only know that fact—or half know it—by rather recondite inference. Light really tells us nothing about its own nature, but only about the superficial aspect of that gross and palpable matter which has interfered with and scattered it before it enters our eye.

Nevertheless the atoms of this solid-seeming flesh and matter as we know it, when analysed into constituents, are turning out to be composed each of a definite grouping of ultra-minute particles, the positive and negative electrons, which themselves hardly occupy any space (save as soldiers occupy a country), and which appear to be of two kinds only—the ultimate indivisible units of positive and negative electricity.

To explain so fundamental a property as inertia is very difficult, and it is doubtful if we shall ever succeed in understanding it except as the fundamental characteristic of substance. We may explain the inertia of matter in terms of something else: we may regard the ether as the only fundamental substance, and all the properties of matter may be reduced to that: but explanation consists in expressing one set of phenomena in terms of something more fundamental, more elementary. In the physical universe that is nothing more elementary, more fundamental, than the ether: so there seems nothing by help of which ether properties can be explained. So far as physics is concerned it would seem that we must content ourselves with ascertaining the properties of ether, explaining everything else in terms of them, and with them be content. To go beyond that would seem to be a problem of Metaphysics and Idealistic Philosophy, upon which no physicist (*qua* physicist) would care to enter.

However this may be, the task of physics proper is still far from accomplishment: in spite of our knowledge, the ether remains a terra incognita; and the reduction of all physical phenomena to manifestations of ether properties is only in process of development. The ultimate nature of gravitation, the ultimate nature of elasticity, the ultimate nature of inertia, the ultimate nature of electric and of magnetic forces, all remain to be ascertained. But if, and whenever, we could reduce them all to an exhibition of the various kinds of motion in one single physical substance, we could all agree that a great stride had been taken, and that physics had in some sort reached its zenith. I foresee a time, perhaps a millennium ahead, when that will be accomplished. Meantime, how far we are from that accomplishment this book may perhaps show.

CHAPTER VII

SUMMARY OF NEW KNOWLEDGE

"One God, one law, one element,
And one far-off divine event,
To which the whole creation moves."
TENNYSON—*In Memoriam.*

I must first ask what is meant by the phrase " the new knowledge." I may summarize what I understand by it, somewhat thus:

(*a*) The fact that electricity is discontinuous, consisting of isolated positive and negative charges called protons and electrons.

(*b*) The electrical constitution of matter, with all that that involves, and the consequent variability of mass or inertia, often wrongly attributed to the theory of relativity.

(*c*) That the atoms of matter form a regular family series, and are all accounted for by groupings or revolving patterns of a fixed number of protons and electrons, on which their chemical power depends. There is a growing tendency to assume that the electric and magnetic properties of the atom are capable of explaining all its chemical behaviour.

(*d*) That the inertia of matter is due to a magnetic field near it, which exists *in vacuo*, and like all magnetic fields is independent of any material concomitant.

58

NEW KNOWLEDGE

(e) That radiation is electromagnetic too, and is only produced by the acceleration or jerky motion of electric charges, so that the clashing together of opposite electric charges is the most powerful source of radiation known.

(f) That radiation is thus only produced discontinuously, so that it is generated and absorbed in units, called *photes* or "photons," each of which must be absorbed or emitted as a whole. There is thus a deep-seated discontinuity, called the quantum, in every relation between ether and matter.

(g) The extensive importance of the velocity of light in the scheme of physics is now emphasised, so that it enters into the expression for locomotion of every combined kind, and into all material or kinetic energy.

(h) That all activity is in the space between the particles of matter, the matter itself being quite inert apart from those fields of force.

(i) According to the most recent theory, every particle is associated with a wave; waves govern the motion of a particle, almost as if they constituted it and were interchangeable with it; in other words, the distinction between waves and particles is getting obliterated.

Many of these ideas seem in themselves to have no connexion with religious doctrines, but a difficulty does arise when they begin to be applied, as they are being brilliantly applied, to cosmogony. Some people have been upset by all this "new knowledge"; but on the whole it appears to me helpful and confirmatory, or, at least, not hostile. Certain speculations about the so-called fate or destiny of the universe are certainly depressing; but, then, they have no philosophic basis; they do not really treat the universe as a whole; they deal with the inorganic or physical universe only, and tell us what seems likely to happen to that. Let us take this point first and clear the ground.

Sir James Jeans, that highly competent mathematical physicist, has applied his great knowledge to the evolution of the solar system, the stars, and the nebulæ, limiting himself specifically to the mathematical and physical aspect of things. He has a plausible account to offer of the removal of the earth and other planets from the sun by tidal action; a theory which at present has replaced the old nebular hypothesis of Kant and Laplace. The formation of spherical bodies from a great cloud of revolving gas still holds good, and is substantiated; but when one goes into detail and reckons the size of the bodies that would thus be formed, one finds they are not planets but suns. The nebulæ are altogether on too big a scale to produce the comparatively small outcome of a solar system. A small nebula would not work in that way. Let me sketch the present form of the nebular theory. A revolving nebula or mass of gas does not shrink in discontinuous rings; the process is rather different. It whirls into an oblate spheroid, the oblateness becoming more and more marked, until it becomes shaped like a double convex lens with a sharp edge, which might come off like a ring, but which, being unstable and liable to tidal action, is much more likely to throw off matter at two opposite poles, in the form of two streamers, which then wrap themselves round the original mass, forming a spiral round a sort of nucleus; like two great spiral arms emanating from the central body and surrounding it. These spiral nebulæ are many of them still in existence, and are seen in various stages of evolution in the heavens, and their appearance strikingly confirms the mathematical theory of their formation. The arms are, however, too long to be stable; they gradually break up into nodules or approximate spheres of gas, which gradually separate further and further from their centre, so as to form a great system of revolving spherical bodies, which ultimately separate out and become more or less independent. It has been found possible to calculate from physical principles the size or mass of these spherical bodies into which the original nebula can break up. They turn out to be much bigger than planets, in fact, to be of the size of stars, that is, of suns. Some of them are much bigger than our sun, and some are smaller; but they average out to about that size. Hence the nebula is not a seat of evolution for a solar system; it is much bigger than that, of a wholly different order of magnitude; and the assemblage to which it gives rise is a constellation of suns, or in its earlier forms a cluster of stars such as are also seen in the heavens. The nebular hypothesis holds true in a modified form, but the result is a constellation. It is a slow and majestic evolution on an extraordinarily large scale that we are witnessing both in the telescope and in the mathematical formulæ.

It is to be noted that the process of evolution is still going on, that the stellar universe is not complete, that suns are still in process of formation. We cannot see the operation actually performed in such a period as our own lifetime. The motions are not really slow, they are so distant and on such a tremendous scale that to our vision they appear stagnant; but we can easily infer that that must only be an appearance. We see the process of evolution at various stages simultaneously, here at one stage, there at another, just as we see the plants growing in a garden. We are in the position of a creature who could only see the garden for some twenty minutes, but who, having first perceived that growth is possible, would realise that he was looking at plants in various stages, some of them in leaf, some in bud, some in flower, and some in fruit, and could surmise that in every case these stages succeed one another; so that their stationary aspect is only subjective. We are witnessing, in fact, the evolutionary process of creation going on.

Our own system of stars is the result of one of these great nebular evolutions, and the particular nebula to which we belong is called the galaxy or Milky Way; the immense size of this can be brought home to us by many illustrations, such as will be found in Sir James Jeans's books. Although so gigantic, it is but a small part of the whole of the universe. Far away in the depths of space the other nebulæ are still revolving, throwing out spiral arms, and producing each its own system of stars. There is nothing in all this to perturb or cause apprehension. Rather it emphasises the majestic scale on which the universe is built. The light which comes to us thence, feeble though it is, brings with it a mass of information which we are learning how to interpret. The spectroscope enables us to analyse the bodies chemically, and to ascertain the nature of the atoms which there exist. We are also able to compute the size of bodies, to weigh them or determine their masses, and to find that the atoms are most of them of the same nature as we experience here, that they vibrate at precisely the same rate, emit light in the same way, and, in fact, are obedient to the laws of physics which we already know and are familiar with. One system of law and order rules throughout all the vast extent. There are many varieties in detail, and some of the elements may be rare there and plentiful here; but they all form part of what we study under the atomic theory, and they fit into our series of elements, each of them numbered and constituted in a certain fairly understood way.

The first lesson thereby taught us, from the point of view of religion, is that the universe is *one*, all of it demonstrative of a single scheme, one system of law and order reigning throughout. Incidentally, we find also that our particular cosmos of stars called the Milky Way, consisting as it does of thousands of millions of suns, is also still revolving; and that our sun as a constituent star is revolving with the rest, round the centre of gravity of the whole, at a rate now estimated at 200 miles a second, carrying with it, of course, its family of planets. That is a detail, but it is confirmatory of the main view. It is a newly perceived, formerly quite unsuspected motion, only discovered during the present century; a motion of which we are entirely unaware by ordinary observation. We have long known that the earth is travelling round the sun at the immense speed of nineteen miles a second. But now we find that the sun itself, though apparently stationary, is travelling at a still greater rate; which shows how little our senses tell us of any kind of locomotion except the locomotion of pieces of matter relative to one another. [It is part of the doctrine of relativity to assert that such relative motion is all that we can possibly ascertain.] So much for the nebulæ, which are so numerous that they can only be counted in thousands of millions. They evolve into suns. Next we may attempt to follow what has happened to a sun.

Our sun is a mass of gas, held together by gravitation, and thus compressed into a mass on the average slightly denser than water; still exceedingly hot, the temperature at its circumference is 6,000° C., while the temperature at its centre is estimated at 40,000,000°, or even more. A hot body like that is radiating energy away; all bodies at a high temperature must radiate, that is, must give out energy to the ether, and so gradually cool down. The sun is in process of cooling down, except that its gravitational shrinkage generates a great deal of heat, and last century was thought to generate all the heat that it emitted. That was Helmholtz's theory for the evolution of heat and light from the sun and other stars. But it was soon found that this shrinkage, or falling together of the materials, is only sufficient to maintain the radiation for a limited period; a period long when compared with human life but short when estimated in terms of the duration of the heavens. Some other source of energy had to be discovered, and this has been put into our hands by the recent discovery about the electrical constitution of matter. It seems that the solar visibility, or radiation energy, depends on the constitution of the atom. The atom is composed of positive and negative electricity, separated from one another, and is, therefore, endowed with tremendous energy. Under certain conditions it is possible for positive and negative electricity to discharge into each other, and give a spark or flash of radiation. The energy thus stored up in an atom ready to be liberated is very great, and when we consider the number of atoms that go to form a visible mass of matter, we find that atomic energy exceeds in amount anything which on the earth has hitherto been experienced. We only attempt to use here the combination of atoms into molecules, a chemical action constituting what we call combustion; to this we owe the heat of our furnaces; an amount of heat utterly insignificant as compared with what could be got by the clashing together of the parts of the atom; though this last appears only to go on under the conditions of pressure and temperature in the interior of stars, but it is sufficient, if it does go on, to account for all their radiation continued during untold æons of time.

Radiation thus generated is of a very high order, higher even than X-rays, but only a small portion of it gets out, most of it is beaten back by having to struggle up through the superincumbent material. The radiation has great energy and exerts great pressure, able to overcome to a large extent the force of gravitation, and this accounts for the enormous size of some of the stars. The waves as they struggle up get lengthened by a known process, until they get down to the rate of vibration or wavelength which is able to excite the eye. We see the stars only by this leakage of radiation, which has thus become visible light. The light itself contains a great deal of energy, which has been produced by the destruction of matter, or rather by the conversion of matter into radiation in the interior of stars. Thus the stars are visible by pouring out continuously this radiation at the expense of their own material. It is now a commonplace to say that the solar energy, upon which we all depend, is produced by the loss of matter from the sun at the rate of 4,000,000 tons every second of time. This applies to all stars, and so the amount of matter in the universe appears continually on the decrease, and the amount of radiation on the increase. Thousands of millions of stars have been emitting radiation for thousands of millions of centuries. One may well ask, where is all that radiation, and what has become of it? We know that radiation is never extinguished, the properties of the ether are perfect, and the vibrations go on with undiminished intensity, spreading out and getting more dilute as the distance increases, but never turning into anything else so far as we know at present. If this process continues, as from the point of view of the physical universe alone it must continue uninterruptedly, it cannot go on for ever. The picture of the universe presented to us is a running down universe; and if all the matter in the universe thus turns into radiation, the result will be a rise of temperature of the whole, of a few degrees above absolute zero; that is, a cold universe consisting of radiation and nothing else. This has been considered a depressing result; for it is as if the majestic frame of things, which we now see in full blast, must ultimately come to an end as a tale that is told, leaving not a wrack behind.

But such a speculation about the fate of the universe is founded upon the data now available to mathematical physicists, who definitely limit themselves in their considerations to the physical universe alone. A prediction like that is an attempt at philosophising, and philosophising with insufficient data. We cannot philosophise unless we take not only the physical universe into account, but the *whole* universe, as General Smuts in his Presidential Address to the British Association and in his book, *Holism*, has been telling us. We cannot understand really the smallest thing unless we take the whole, every aspect of it, into account. A theory of what is to happen to the universe as a whole is impossible for us with our present knowledge. Mathematical physicists can trace the fate of the material universe so far as the data allow. If they limit themselves to purely inorganic data they arrive at the conclusion I have indicated. Yet conspicuously the universe contains far more than matter and energy, it contains a number of things which physics does not take note of, and which if taken into account may upset the conclusions and render them quite futile.

Moreover, even in physics alone there are many discoveries in the womb of the future still waiting to be made. We have only recently discovered that matter turns into energy. Why should we not, in the course of years, discover that energy can turn into matter, that the process is reversible, and that we have taken only a one-sided view of it? There are many facts now beginning to be known which hint at such a process, that is, at the reconstitution of matter out of radiation. Whether this is ever discovered or not, I would not build upon it too energetically. My belief is that it will be found that the running down of the universe is only a human conception, and that it need not really come to an end in the dreary way now imagined. If that should be the end, it must mean that the universe had a beginning, else it would have run down already. The fact is we are not entitled to speculate with our present knowledge upon these tremendous themes. One cannot imagine a time before the physical universe came into being, nor can we imagine a time after which it will have ended. But what I want to insist is that even if these temporary deductions are substantiated, and if it be found that matter is, after all, an ephemeral phenomenon, which having gone through a process of evolution will come to an end, still we need not be depressed. For the amount of matter in the universe is after all very insignificant as compared with the extent of space. The heavenly bodies are at an enormous distance from each other. The things that should really attract our attention are not the particles of matter, but the properties of the space between them. And when we turn to that, all our ideas are modified. We gradually find what the true function of matter is.

Here and there, or at any rate here, a sun has been deformed by tidal action till it shot out a number of planets, which cooled down rapidly, till one of them stays at a temperature near 300° above absolute zero, so that atoms combine into complex molecules and liquid water can exist; under these exceptional circumstances life has somehow managed to get hold of that matter and animate it. Incarnate life requires very narrow limits of temperature for its existence, and is impossible in connection with most of the matter in the universe; life only makes use of matter in the rare case when these conditions are fulfilled. We only know of life when it is incarnate in matter, but that is due to the limitation of our senses.

MY PHILOSOPHY.—PART I

We do not see life or mind directly, we only study them by observation of an organism which is animated by them, that is by what we call its "behaviour." All we see of each other is the bodily organism, all the rest is an inference. We cannot tell another person's thoughts except by his actions, or by the vibrations which he may choose to emit. I do not say that this is inevitably so, but it is so under our present circumstances, and supplied only with our present senses. If our senses do not perceive matter, they perceive nothing. We live in a mysterious universe, full of all manner of things of which we are conscious, conscious in our own selves. We appreciate colour and beauty; but colour after all is only vibrations—that is all we get from a coloured object; we interpret the rapidity of vibration as colour. Colour and beauty are interpretations of the mind. So are art and music and literature: they are all interpretations. They can be recorded or incarnate in matter, which we then call a work of art; but the work of art in itself consists only of pigments on canvas or black marks on a bit of paper. All the reality of these things is in the region of the unseen, the unsensed; they are all mental interpretations.

Our sense organs, including the brain, only give us indications of the physical objects around. The realities underlying them we have to infer; and we do this, not by the brain, but by the mind. It is *yourself* that admires a landscape or a work of art; it is yourself that has the feelings of beauty and design; aye, and it is *you* that have the aspirations and the hope and the love of which you are conscious, not any material organism or any part of that organism. The organism we have ourselves constructed, and we use it for a time. Matter seems to have an ephemeral purpose; it does not last very long, it wears out and decays, and ceases to be useful; then we discard it. But it never was a part of

68

ourselves, it was the instrument we used for manifest-ation, for making signs to our fellows, and for getting signs from them; its function is to display, to manifest, what by all analogy is not in matter at all but in space.

What life and mind are I do not pretend to say; but I know that they are not functions of matter. We employ matter in the exercise of our functions at present; but there is every reason to believe that we ourselves continue to operate even apart from matter, and that the destruction of the material organism only interferes with our mode of manifestation. Things which are not associated with matter have no obvious effect upon us, at least to our present senses. They are outside our ken, and we are apt to imagine that they do not exist. The Behaviourists are physiologists and pathologists who study the behaviour of an organism, its reaction to drugs, the way its own gland-secretions act upon it, sending the organising substances called hormones to every part. They study these things with minuteness and skill, and their results are well worthy of our attention, so long as we remember that they are telling us about what they know and can observe. But we cannot trust them to philosophise; they are attending to a small part of the universe, a merely material part, and we cannot philosophise on a part only.

The business of science as now understood is mainly concerned with mechanism, some time ago we might have said with material mechanism, and that is what chemists and most biologists are still mainly concerned with. They still very often limit themselves to a study of the structure and behaviour of organisms; all which is in the main true, but by no means the whole truth or the most important truth. The physicist has gone beyond material mechanism, he deals with radiation and many etheric phenomena; and now, under the influence of Faraday and Maxwell and Einstein and other great philosophers, is more concerned with the phenomena that occur in space, or in what may be called the ether.

A physical theory is by no means complete when it describes the behaviour of matter alone; it is constantly referring to the fields in space, wherein all the energy and activity really lie. It teaches us that matter is inert, takes the path of least resistance, and obeys the forces acting on it with perfect accuracy; but it has no spontaneity, no real energy, it is the index and the sign of something which eludes our senses, but which we can infer by its aid.

If, then, we find in ourselves, or have reason to infer in the universe, things which are only partly displayed by matter and are still largely concealed, we need not deny their existence because they are outside our sensible ken. We may feel assured that they are realities, the ultimate realities of existence, and that they will last much longer than any material mode of displaying them can possibly last.

CHAPTER VIII

MACHINERY OF GUIDANCE

"And of the body I think as the machinery
of our terrestrial life evolving towards conscience
in the Ring of Reality; and thence of the mind
as that evolved conscience."

ROBERT BRIDGES—*The Testament of Beauty.*

Among the properties of matter which have been
observed from the most ancient times is one to which
we must now make more adequate reference. Certain
pieces of matter were endowed with what seemed to be
a real spontaneity. They were not merely pushed
about, like the inorganic variety, they seemed to move
of their own volition. They were not merely pushed
from behind, they seemed to have the power of
anticipation. Their locomotion seemed to have something purposive in it; that is, it was directed towards
something in the future. They not merely avoided
dangers in the present, but they sought for food in the
future. They did many other things of a kind impossible to inorganic matter; they fed and grew and
reproduced, and handed on their activity to generations
ahead. These pieces of matter were called animated.
The nature of animation was unknown, and is unknown
to this day, but it was something that had to be investigated by aid of the behaviour of the matter affected.
We are bound to study the behaviour of material bodies,
and the properties of animated matter are now by
bio-chemists and bio-physicists being carefully observed,
especially in some of their simpler and more instructive

71

forms. Experimental bio-physics and bio-chemistry are being pursued successfully at Cambridge, as well as in America and Europe. The Strangeways Laboratory at Cambridge is an admirable institution, which I have visited, where the properties of live tissue are studied in their simplest form.

The properties are very peculiar. There seems to be a certain size and shape appropriate to each live organism, although it is made up of a great number of cells, each cell having a kind of individual life of its own. So much so that if it is damaged, as by a wound or amputation, the organism tends to restore itself and replace the missing parts, with more or less success: as if there were some standard or norm which was aimed at and to which it tended to return. It is familiar that cuts heal, even in the highest organisms; in the lower the recuperative processes are still more remarkable. If you cut off the leg of a newt the animal soon reproduces a new limb. If you tear a starfish in half, each portion replaces what is missing, and the result is two starfish instead of one. The structure of an organism does not depend on the food taken in, but on some controlling or directing principle, which is what we call animation or life. Any wholesome food is utilised by the organism, and is built discriminatively into its various parts.

Now let us think how this specific organisation can be brought about. We had already found particles behaving in a curious way, and the causes of this behaviour we ultimately associated with gravitation or cohesion or with what we called an electric or a magnetic field near them; so that the behaviour of the particles was an immediate consequence of those fields in space. I would press the analogy to the behaviour of animated particles too. The organism is an assemblage of inert particles; and yet it behaves in a certain complicated fashion, owing to the interaction with them of some unknown entity in space which may be briefly designated as "Life." It is this interaction of life with matter which causes and sustains animation, and must be the ultimate explanation of biological processes.

Animated matter behaves very differently from inorganic material. It not only has a characteristic size and shape, but it acts spontaneously, and adapts itself to circumstances in a way no piece of mechanism does. A clock-work mechanism can be made to imitate the motions of a singing bird or a kitten, but it always does the same things, and is not affected by occurrences in its neighbourhood. The live thing also is self-moving, but it is aware of what goes on round it, and can modify its behaviour accordingly. The differences between the gambols of a dead leaf and the gambols of a kitten are obvious, but when analysed are not hopelessly discordant. The kitten's body is self-moving like clock-work, but the propelling force even in clock-work is really external to the matter. Thus if the motive power is derived from a raised weight, a gravitational field in space contains the stored energy; if the motive power is derived from a strained spring, the energy is in the elastic medium which unites the molecules of the spring by what is known as cohesion. Neither the atoms nor the molecules of a strained spring are themselves strained, they are merely altered in position, — all that we can do to matter is to move it,—it is the connecting mechanism which is subject to strain, and so tends to bring the molecules back into their original configuration. The motive power is always outside the actual particles of matter, which in every case are absolutely inert.

No difference is to be expected between a molecule which is part of an animated cell and any other molecule:

73

all are activated by outside agencies. We imagine we understand the agency of gravitation and cohesion more than we do the agency of life, but in actual fact we are fairly ignorant about all. There is a difference however. In a gravitational or an electric or any other field, energy is stored and expended. Those fields belong to the physical universe, and their energy has to be taken into account. In life apparently there is no specific energy. Life is not energy, it merely directs the energy which it finds available. All the energy relations of an animated body can be dealt with in physics just like that of any other mechanism. No energy is added to a body by the fact that it is alive. Sir Arthur Keith has recently spoken of life as a form of energy. I challenge him for any evidence of such a statement. Were it true, life would be convertible into other forms, and would have a mechanical or thermal equivalent. The phrase "vital energy" has been used, but to a physicist it is an erroneous one. "Energy" is not a vague, it is a very definite, term. People seem to think we are using terms haphazard. We are not. The element which life adds is not energy but guidance, a power of directing energy into channels which otherwise might be unoccupied. Thus it is that various structures are due to the agency of life, from a bird's nest to a cathedral. The mechanism whereby the stones were lifted is wholly in the province of the engineer; a living architect is responsible for the arrangement; but beauty consumes no more energy than does ugliness.

Life makes no direct appeal to the senses, and is therefore an unknown entity. We only study its properties by the behaviour of the organism thereby animated; and to seek to explain those properties in terms of, or as nothing more than, the movements of inert matter, would be to throw away all our analogies and return to the physics of the eighteenth century.

LIFE AND GUIDANCE

We know that as far as physics and chemistry are concerned the atoms of matter in an animated organism differ in no respect from those in an unanimated one; that is they are just the same in things which are not organisms at all, but merely accidental collocations of material, like a stone or a chair, of which pieces can be chipped off without the slightest attempt at renovation. (Parenthetically, I am not so sure about a chair. That is not an organism but is a product of one. It does correspond to an idea. It has a shape of its own, and if slightly damaged or a leg broken off it is very apt to be mended and its part restored by the agency of life.)

We need not suppose that life modifies the energy of an organism in any respect. All the laws of physics are completely obeyed by it, and no exceptions occur. The energy displayed can be accounted for in terms of chemistry, just like the energy of a steam-engine. The organism is in fact a machine like any other machine. In so far as it differs it differs not by rebellion or exception, but by supplement. It moves with a purpose. So it is also with the musical instrument called an organ. The energy is all supplied by the bellows indiscriminately: all that the executive musician has to do is to direct it into desired channels. All these instruments have superadded to them a something which, though it modifies the amount of energy in no particular, yet guides and directs it towards the attainment of some *end;* a thing never observed in inorganic matter. An organism in fact need no longer follow the path of least resistance; it can decide to ascend a mountain, or fly to Australia, or to do some other feat of endurance. It can direct its energy towards the attainment of some object, the overcoming of some difficulty. The foreshadowing of possible complications is within its scope; it has a sort of *aim.* It is not only actuated by experience

75

of the present, it has memory of the past and some anticipation of the future; and its course accordingly is directed: but always without any breach of mechanical laws. The behaviour of animated matter is just as obedient to those laws as that of any other kind, but it is supplemented by a controlling guiding or directing principle,—a principle which in ordinary physics and chemistry has no existence, and is never taken into account.

This organising principle is fully recognised by the philosophic biologist or biological philosopher Professor Driesch, of Leipzig, and he applies to it the Aristotelean term "entelechy," a rather complicated though historical name for an organiser or arranger or completer with some end in view. I prefer to use the simple term "life," for entelechy corresponds to the characteristic mode of action or manner in which life enters into relation with matter. In an article in *Nature* for October 25th, 1930, Professor MacBride, the eminent biologist of the Royal College of Science, S. Kensington, under the head of "The Problem of Epigenesis" reviews three German books on Embryology, and in so doing surveys the ground in an interesting manner. I will make one extract from his article, on page 641, which speaks for itself, and is entirely in accord with my own view.

"What are we to say of the experiment recorded by Dürken in which the tail bud of one newt embryo was grafted into the body of another near its forelimb and developed into a new limb? Presumably the cytoplasm of the tail bud was 'organised' so as to produce the tissues of an adult tail. How then was this organisation so completely changed as to produce a limb instead? No wonder that Dürken says that in cases like this, physical and chemical explanations leave us completely in the lurch, and we must have recourse to the conception of the 'biological field,' an influence not in the living matter itself, but in the space, presumably the ether, around it."

LIFE AND GUIDANCE

I commend this notion of a bio-field as an addition to our list of gravitational, electric, and magnetic fields. It so happens that we have special facilities for observing the actions of organisms, since we ourselves are each endowed with a machine of that kind; moreover, life in us, having attained in the course of ages a higher grade, has blossomed into mind and intelligence, so that we can study our behaviour with some hope of understanding it, and some hope of access to the controlling power which determines our actions. In this effort we are justified by the experience of our scientific ancestry; for we have always found that when a thing acts mysteriously, we find a cause in space which shall explain it. It is natural to seek for such a cause now. At present, when called on to explain a spontaneously jumping bean, for instance, we can only say that something in it is alive, and stop there. We never seek to explain the activity of live things beyond suggesting that they are alive. We may never reach ultimate explanation, but we are always seeking to carry on to the next stage, one step in advance.

So it was when we postulated a gravitational field to explain the motion of the planets; we arrived at nothing ultimate, but we took a step in advance. So also when we postulated electric and magnetic fields existing in space; especially when Maxwell made the important discovery that those two fields in combination would explain radiation, and thus ultimately rationalise all the actions between the isolated particles of matter, including their chemical action. For it is probable that no piece of matter acts on any other piece directly, but always through the intervention of the properties of whatever it is in space that forms the intermediate link or connexion between the particles. So now, when we seek for an agency that shall not interfere with amount of energy, shall leave that constant, but shall guide and

77

direct the motion of particles, we must think whether any such agency has been either discovered or foreshadowed in the present state of physics.

And here we have to make use of the wave theory of matter. Modern physics has shown that, like the corpuscles of Newton, every particle has associated with it something periodic. The electron does not now appeal to us as a minute spherical charge and nothing else, it has been found to have something associated with it, a series of waves. In fact it is found that many of the properties of a particle of matter can be possessed by what is known as a "group wave"; in other words that the energy of a particle can be expressed as the energy of a set of group waves, and that these, strangely enough, obey the laws of dynamics. Consequently it is realised that the particle and the wave are much more united than ever they have been before. A wave may exist without a particle. A particle can hardly exist without a wave. The waves seem the most fundamental things.

The immaterial waves which are thus indicated, and of which a localised and identifiable group imitates the behaviour of a particle, have not as yet been properly or fully discovered. Something is known about their wavelength, and something about the speed at which they travel, which is quicker than the velocity of light. The waves last mentioned are mere forms that convey or transmit no energy, they are in fact not effectively energetic, or so to speak real *things;* they are more geometrical, they are called "form waves." Their progress is not like that of a material entity; and yet they are supposed to guide the particle to its destination. In other words they exert a controlling and directing influence without imparting any energy. They may be said very roughly to act something like the rails which guide a train.

LIFE AND GUIDANCE

We know too little about these waves to surmise how it is that they have any anticipation of the future, or can direct the particle along paths beneficial to the organism. All we know is that they have the power of direction, of determining a path, and that they do not interfere with amount of energy. They thus have some of the attributes which we have been accustomed to associate with life, or, I should say rather, with life and mind. For mind certainly is a controlling and directing influence.

It may be said, and it is probable enough, that mind in itself belongs to the psychic aspect of the universe, and is not to be explained physically. In that case the form waves, if they have any mental or biological significance, must be only the physical agent utilised by mind, enabling it to get into touch with matter. How a psychic thing can act on a physical thing has always been a puzzle, and some unification of physics and psychics has always been sought. But if matter can be expressed as group waves, the connexion of a guiding principle with form waves becomes intelligible. We are now beginning or seeming to begin to get at something physical which acts in a psychic way; and it is an advantage not to be lightly thrown over or discarded. It may lead to nothing, or it may be a step in the right direction. It is not likely to be an ultimate step; for after all what we think we have found is only an intermediary of a physical nature.

My hypothesis is that these form-waves constitute the physical mechanism whereby life and mind operate on and direct material particles. The nature of life and mind is still unknown, and it is probably not for a physicist to attempt a solution. It may be even doubtful at present whether these form waves themselves actually exist: it may be that they only represent probability, and are only another will-o'-the-wisp born out of mathematical calculations and speculation thereupon. They are represented by the symbol ψ or its square, and are treated symbolically. To treat them as physical realities is equivalent to a hypothesis. But the way to examine a hypothesis is to give it attention, trace its consequences, and find in it some flaw. Hitherto the nature of life and mind has been mysterious, and when we are involved in mystery any clue may be welcomed and followed. That is all I claim at present.

What I am sure of is that no explanation will be obtained by a study of material processes alone. The particles of a brain are, like any other particles, inert. If they operate so as to achieve results, it must be because they are acted upon by something; something which controls the particles; something which can be expelled from them and whose interaction can be interfered with, but something of whose activities, their movements and behaviour, are the sign and index or demonstration, and is otherwise unknown. Mental processes are demonstrated by the interaction of mind with matter. If the particles are damaged, the demonstration is interfered with or suspended. Mind itself is not necessarily damaged by damage to the brain. A hatchet used on a loud-speaker will interrupt its flow of words, but will have no influence on the etheric cause of that linguistic flow. Only the machinery is damaged; and accordingly the source, or mind, can no longer operate so as to appeal to our senses. If we cannot observe particles of matter we can observe nothing; they are our only means of observation. The whole of the real activities are in another, a super-sensuous region, and are known to us only by mental inference.

Note that even colour is a mental inference. There is nothing in external nature but rates of vibration: we interpret them as colour. So also we have to

LIFE AND GUIDANCE

interpret a picture or a piece of music: physically such things are insignificant. What is a poem physically? Black marks on a bit of paper. The universe as we perceive it is largely our own interpretation: though our faculties doubtless depend on our sense organs and on the information they have given. We have gone a long way beyond them in aesthetic interpretation, but fundamentally we are limited in our perceptions of reality. What Reality actually is, and how it might appeal to a different order of being, we can make no pretence to know. We can only grope along from step to step, and hope that, as far as we go, we are going right; though we are well aware that we are only animated organisms in a great and mysterious universe, the full meaning of which we can only surmise by faith.

Appendix to Chapter XIII

When this Cambridge Lecture was published in *The Hibbert Journal*, the Editor rightly asked for further elucidation about guidance without expenditure of energy. And my reply to him was as follows:—

"There must be a difficulty about guidance without expenditure of energy, because many people have felt the difficulty: Arthur Balfour, for instance, always boggled at that. But I think his objection was rather deeper seated, in that he could not see how mind interacted with matter at all, whether to impart energy or to change the momentum. No doubt he admitted the fact, but could not see the mechanism; nor can I fully. But what I see is that the two functions, the doing of work and the effecting of guidance, are separate and distinct; yet both mean the exertion of force. The rail analogy isn't a bad one. The rails exert force on the train, but they do not propel it one whit. You may say they retard it; but that is mere accident, due to the imperfection of matter; there is friction, and the rails wear out; but there would be no wearing out of etheric rails nor any friction. Tennyson adopted the analogy when he said:

'Let the great world spin for ever
Down the ringing grooves of change,'

MY PHILOSOPHY.—PART I

"And an Oxford undergraduate expressed something of the sort after a course of lectures on determinism:

'Damn! it is borne in upon me that I am
A creature that moves
In predestinate grooves,
I'm not even a bus, I'm a tram!'

"Anyhow the rails bring the train to its destination. Of course, the real guidance was exerted when the rails were laid down or rather by the mind that planned the route to be taken. That mind exerted no force on the trains; but it influences them to this day. And the mechanism is fairly clear throughout. The train merely has to take the path of least resistance, just as a planet does on the Einstein theory of the curvature of space near a large body: not exactly curvature in the technical sense, as when we speak of the finite universe and the guidance of a ray of light round and round, but still a modification of the Ether of Space, so that an ellipse is the easiest path for a planet to take.

"If we press the analogy in order to assist in contemplating the actions of live things, some vagueness is inevitable; yet I feel that life affects the balance of energy no whit: it does not interfere with its conservation: it merely controls the path of the particles. The physicists object that though that is true, and the conservation of energy is still intact, the conservation of momentum is rather modified, for when a thing is deflected by a force at right angles, its momentum is changed, not in magnitude, but in direction; and this is equivalent to the imparting of an additional momentum at right angles, in accordance with Newton's Second Law of Motion. That is where the hitch comes in at present. We have to find some machinery whereby this can be done. I have dealt with this in my book *Beyond Physics*. The wave theory of matter is a recent discovery. It involves the perception that a particle is closely associated, and in some sense identical, with a set of waves—'group waves' as they are called—which convey energy; and that these group waves are directed by another hypothetical set of form waves, called Ψ by Schrödinger, which, though they possess energy, transmit none, but nevertheless guide the group waves and determine their path. They do this in geometrical fashion, not involving any energy. They do not exert force on the group waves, either lateral or any other. The group waves are a sort of offspring or immediate consequences of the form waves. They inevitably move as determined by their constitution. And the particle, in so far as there is a separate particle, has to go wherever the waves take it.

"All this is quite recent, and the only question is whether the form waves have any real physical existence, or whether they are nothing more than a mathematical abstraction. That is why Eddington put up his notice, 'No admission to philosophers during the alterations.' And that is why my ground is not quite clear. There is much more to be discovered, but my instinct leads me to speculate beyond actual discovery, in view of the fact that here we perceive for the first time a guiding and directing agency which achieves its end without interfering with energy. That on the one hand. On the other, the undoubted facts of experience that life and mind are just in this predicament, and have been doing this same thing all the time, without our having any idea of how they do it."

PART TWO

EVIDENCE FOR AND CONTROVERSIES CONCERNING THE ETHER

CHAPTER IX

MATTER, ENERGY, AND THE ETHER

When from time immemorial men and animals exercise their senses, look round and take notice of the physical world, that is of what they call the universe, what is it that they apprehend? First they apprehend matter in solid or liquid form, this without exception; and they appreciate it initially by its resistance, that is by their sense of force or muscular sense, bumping up against objects and learning to avoid them. Next they similarly feel the resistance of the air, and so arrive, some time later, at the notion of matter in the gaseous state, which may be said to involve some element of discovery, not yet made by small children and animals. After this there was a long interval, until some of the Ancients became impressed with the space between the pieces of matter, and so began to emphasise the importance of what they called "the void." They found that the universe consisted of portions of matter in its various forms, separated by empty space, which at first and for the generality of mankind still seemed to be mere interruptions or gaps in the continuity of matter. Matter still seems to us the important thing because it is for that that we have special senses. We realise that it has certain properties which we can easily explore, and only the most philosophical of us go beyond matter and enquire whether empty space or vacuum has any properties. The properties of vacuum certainly make

87

no direct appeal to our sense organs, and must be entirely a subject of inference. Still in comparatively early days it was found that certain things went on there; that light, for instance, was transmitted by, or at any rate existed in, empty space. Light is generated by matter and absorbed by matter and gives us information about matter at a distance; but for a long time it told us nothing about what conveyed it, or indeed whether it was a thing that was conveyed, whether it was a thing that travelled with any velocity, or whether it merely existed simultaneously as a connection between the particles of matter.

Then in the early nineteenth century it was found that light had a definite speed of travel, which was the first information obtainable about the properties of space. Light was perceived to be an affection of space, not of matter. It travelled at a tremendous rate that no matter could hope to compete with, a million times quicker than sound,—sound being the quickest kind of effect transmitted by matter. Light at first was thought to be corpuscles shot out from bodies and flying at an immense speed quite freely, until they encountered some other body of matter, when they produced some visible or tangible or otherwise detectable effect; the first effect noticed being that they stimulated our sense-organ the eye whenever they entered it, and thus gave us information, not about the corpuscles nor about the medium through which they travelled, but about the material object from which they came.

The eye had the property of analysing this impression into different kinds of colour; but when this analysis was carried further, as by Newton it was by means of a prism, it was found that there were many kinds of light that did not affect the eye, though some kinds, the ultra-violet, affected photographic plates, and all kinds, especially the infra-red, generated heat in the bodies which absorbed them. Then, by more refined observa-tion, and by studying the theory of shadows, it was found that light had a *periodic* property, that it involved a process that repeated itself rapidly at regular intervals; thus causing fringes and rings and bands and other interference phenomena, which could only be accounted for by something periodic in space and time and there-fore akin to the periodic progressive phenomenon called a "wave."

And so at the beginning of the nineteenth century the wave theory of light was born, not indeed wholly for the first time, for Newton had observed some of these phenomena, and had attempted to account for them by endowing his corpuscles with some periodic or revolving character. But the wave theory seemed at first to abolish corpuscles altogether, it took their place and did instead; it was thought that light consisted of waves travelling in a medium which fills space,—this medium being possessed of the properties necessary for the transmission of waves. Accordingly a further inference was made; the medium which fills space was discovered and called the ether.

Many hypotheses were made concerning this ether, and right down to the times of Lord Kelvin it was considered an exceedingly rarefied kind of matter, and many attempts were made to investigate its properties experimentally, all of which ended in failure, so that some philosophers at the present day find themselves able to doubt whether such a medium actually exists, and some few express their doubts scornfully. Everyone, however, admits space, and time, as the two essentials to a periodic movement, of whatever kind it might be ; and accordingly it is now the fashion to speak of space-time as the representative of the ancient void or vacuum, without necessarily supposing it to be a substantial reality, and yet admitting that it has physical properties, which we can hope to investigate by experiment.

There is some doubt whether the ether ought to be called a substance: it differs from every substance so far known to us, yet it is very fundamental and therefore substantial. No one now supposes that the ether is a rarefied form of matter, or that its properties can be expressed in terms of mechanism or material behaviour. It is evidently something more fundamental than matter, something of whose properties we have very little knowledge, but still something of which the importance is recognised, so that the attention of physicists is more and more directed rather specially to it.

The universe is now regarded not so much as an assemblage of material molecules interrupted by empty space, but rather as a great region of space-time interrupted here and there by a particle of matter; for it has now been found that matter is essentially discontinuous, and on the average rare; that it consists of minute particles separated from one another by considerable intervals, only joined up and connected into a coherent whole by the space in which they are immersed. And instead of explaining space in terms of matter, the attempt is now being made to explain matter in terms of the space in which it exists, a study which has not made very much progress, though an immense amount has been discovered. Some indeed think it probable that I might add "and of which it subsists," being impressed with the idea that the elements of matter have as it were crystallised out of the ether, so as to affect our sense of touch and the other senses.

Physicists have been helped in this quest by the incidental discovery of some further properties, which are now found to be really properties of space, namely the electric and magnetic fields, which to the Ancients were remotely known in the form of affections of special kinds of matter, such as amber, which when rubbed attracted light bodies, and the iron-ore called lodestone, which exerted a specific attraction on bits of iron. These incipient phenomena under careful scrutiny have shown themselves not to be material properties at all, but properties of the vacuous space in which the bodies existed.

Faraday began this attention to space in connexion with electricity, directing our attention from the charged bodies to the space surrounding them, and showing how the behaviour of the bodies could be modified by varying that space. Conductors were only the boundary of an electric field outside them. He also mapped out the conditions of a magnetic field by the use of iron filings, which grouped themselves into certain patterns, illustrating the stresses and strains which occurred in the region, the vacuous region, near a magnet. He used filings as an indicator or demonstration of the magnetic field, which field was really independent of any particular mode of observing its effects.

This extension of the idea of the properties of space has gone on developing, so that when an electric current was discovered—which again was only in the nineteenth century—it was gradually perceived and taught that the electric energy thus transmitted by a conductor was really conveyed by the space round it—that it did not travel by means of the conductor—and that the conductor directed the path of the energy merely by dissipating such portions as came into it from outside. (See Poynting, *Phil. Trans.* Roy. Soc., 1884, p. 343.) The rate at which a telegraphic signal could be conveyed was not dependent on the matter of the conductor, but was a property of the space around it, that the electric impulse in fact travelled with the speed of light; so that at one time it was uncertain whether anything really travelled along the conductor at all.

Maxwell himself seemed doubtful about this longitudinal flow.

Then, at the end of the nineteenth century, it was found, surprisingly, that electricity too was essentially discontinuous, like matter, and that there were small particles of electricity called electrons, which really were impelled along a conductor conveying a current. But the rate at which those particles moved was not known in most cases, and in the few cases when it became known it was found to be exceedingly slow, much slower even than sound, that it could rather be called a creep or a crawl than the instantaneous flash with which the energy really travelled to the distant station.

So once more it was found that in electricity, in magnetism, and in light, the really effective medium, whatever it was, existed in space-time, and that the movements of the particles of matter were only an index, a demonstration, a phenomenon which could be observed; it was found that the perceptible motions of matter were consequent upon the real phenomenon, which was operative in that which appealed to our senses as empty space, or in other words did not appeal at all. And so gradually we have come to the conclusion that the atoms of matter and of electricity have no motive power of their own, but that they are obedient to the resultant forces that act upon them; in other words that they take the path of least resistance, and simply drift wheresoever they are urged. True, the urging may be due to or in consequence of the existence near them of other masses of matter, but the urging is done by a machinery which is not mechanical, but rather what I should call etherial, that is to say by a state of the space or medium in which the particles find themselves. The whole of the activity therefore is a matter of inference, not of direct perception. All that we are able to perceive is the motion of the particles, but these are not primary, they are secondary results of some primary activity which makes no appeal to the senses at all.

Some conception of this revolution in ideas was dimly grasped by Newton in connexion with his theory of astronomy, or rather by Newton and his successors. The very existence of matter was found to endow it with a singular property of appearing to attract all other pieces of matter, so that for the most part they revolved in orbits round each other exactly as if they were mutually attracted. This attraction might be due to an electric or a magnetic or a gravitational field, but anyhow it was not action at a distance, it was due to something which existed in space and tended to drive the pieces of matter together. Only their motion enabled them to keep apart. In so far as they moved they were supposed to be endowed with energy, and all through the nineteenth century this energy seemed to belong to the particles; until in the twentieth century Einstein suggested that even this property did not really belong to them, but was the sign and symbol of an energy which really resided in space, a far greater energy than anything that had yet been discovered, and of which the locomotion of the particle was only a faint residual outcome. The real outstanding absolute speed was the velocity of light. All else was relative. The motion or momentum of a particle could be expressed in terms of the space near it. The electric theory of matter had already shown that the inertia or mass of the particle did not really belong to it, and was in fact not constant, but could increase to any desired extent when it moved with sufficient rapidity through the medium; and further that the energy of the whole motion was representable as the energy of this small increase in mass moving with the velocity of light. The expression for the momentum mv of a particle is $p = \sqrt{m^2 - m_0^2} \cdot c$; that is a minute mass moving with the velocity of light.[1] So that this velocity c comes in every time. The mass of the particle did not belong inalienably to that particle, but was due to the properties of the space near it, and was modified by the conditions existing in that space. In other words the particle had no energy at all, space contained the whole of the energy, and the motion of the particle was only an indication or index or demonstration of the remarkable properties of the space near it under different conditions.

But these conditions we have no means of ascertaining except by observing the particle, and accordingly it requires an effort to realise them; that is, to realise that therein is the real phenomenon underlying even so simple seeming a thing as the locomotion of a piece of matter. Locomotion is the only thing which our muscles enable us to bring about, and it therefore constitutes the means whereby we explore the universe. When we see the needle of a galvanometer deflect we say that it is the sign or symbol of the unknown entity called an electric current in its neighbourhood; just as when we see a dead leaf careering about and apparently amusing itself, we know that it is the sign or symbol of an air vortex, which is the real phenomenon demonstrated by the motion of the leaf. When we see a tree bending and moving about, its branches waving, we take it as the index of a gale of wind. The clouds drift about in the viewless air; the real happenings only make appeal to the senses by their operation on the inert

---

[1] It may be well to show how this follows at once from the relativity expression for the mass of moving matter whose mass at rest is m_0

$$m = m_0 \left(1 - \frac{v^2}{c^2}\right)^{-\frac{1}{2}}$$

$$m^2 \left(1 - \frac{v^2}{c^2}\right) = m_0^2$$

$$(m^2 - m_0^2)c^2 = m^2 v^2$$

which instantly reduces to the equation in the text.

MATTER, ENERGY, AND THE ETHER

portions of matter subject to their influence. The moral
of it all is that when we seek the real causes of things,
which is the object of science, we must turn our attention
away from the pointer-readings and other index demonstrations, and concentrate on that which cannot be
directly observed, but which can only be inferred by
the action of the mind.

CHAPTER X

THE ETHER AND THE FORMS OF ENERGY

Our conception is that the physical universe consists of one substance and only one, though in many modifications, which enable us to perceive it under various aspects. Experience shows that the Universe as a whole is greater than the physical universe, inasmuch as it contains life and mind and other entities which can hardly be called "substance." These entities lie beyond the scope of physics: they make use of substance and manifest themselves thereby. If, for brevity, I hereafter use the term "universe" without qualification, I shall mean the physical universe; for it is that in which we live and are active, it is that which scientific men explore, and it is about that that we seek to know. For brevity we will call the one fundamental substance the ether; it is co-extensive with Space.

Most of the ether is free and unmodified, and apparently infinite (in the sense of having no boundary), a uniform substantial reality, having certain properties, which it is legitimate to investigate, and about which we may hope to come to some conclusion. What we at present know about the free unmodified ether is very little; its constitution is unknown, and probably the laws of Dynamics, as at present understood, do not apply to it. Dynamics has to do with the application of force and the effects thereby caused: we cannot apply force to the ether, and therefore we cannot examine the effects of force upon it. It appears to be incapable of locomotion, and may be taken as our standard of reference. Any motion taking place in the ether may be called absolute motion—unless in due time we make further discoveries and have to modify that view. At present we have no means of ascertaining absolute motion even in this sense: whether discoverable or not, we define it as motion relative to the ether, that is to the main bulk of unmodified ether.

WHAT WE KNOW ABOUT FUNDAMENTAL THINGS

The things we know about the great bulk of the ether may be summarised thus: It is not a generator of force, nor is it subject to force; but it is a transmitter of force. *Transmission* is its main function; and we have learnt that it transmits force at a certain definite and measurable pace, a fundamental velocity, which is called "the velocity of light,"—a fact which shows that it belongs to the physical universe, and has finite properties which we may hope to explore. It makes no appeal to our senses, but it is not beyond our ken.

If the whole of it were of this free and unmodified kind, we should know nothing about it. It only comes within our ken because scattered about it here and there are certain modified portions, isolated from each other, but not independent of each other, nor yet independent of the substance of which they consist. Through it these modified portions exert force upon each other, group themselves into patterns, and excite vibrations which travel through the ether at the one measurable pace. The patterns into which they group themselves are known as the atoms of matter: these do appeal to our senses, for of them our sense organs and the whole of our present bodies are composed. We can touch and

G 97

handle these portions of the ether. Their behaviour is the theme of all physical science: they exert force on each other, they are susceptible of locomotion, and the laws of this motion are known. They are subject to the laws of dynamics, which they perfectly obey: in other words the laws of dynamics are a concise summary of what we have discovered about their forces and motions.

Force and motion are our two primary sensations: and the main business of Physics has been to explain all observable phenomena in terms of these two things. We apprehend force through our muscles, that is through our muscular sense: and we encounter force by reason of our power of movement. It is through locomotion that we encounter resistance, and thus gain an impression of physical objects with reference to which we move. The only motion we are aware of is this relative motion, the motion of one set or group of particles in relation to other groups. We apprehend motion too through our muscular sense, and also, either directly or indirectly, through our optical sense. We have a sense also of rapidity of motion and of the duration of motion: we have a sense of exertion and of fatigue. By this means we have acquired the idea of Time, and also of Space. Space, time, and matter seem to us the three fundamental abstractions, in terms of which we hope to summarise what we know of the universe. Matter appeals to us through its resistance, its obstruction, muscular action impeded, static force. Space appeals to us through muscular action free, room to move about, interval between one material object and another. Our bodies give us a standard of size and reach: objects which we can reach by stretching are comparatively near, objects which we can only attain by a long walk are distant. Time appeals to us both through our sense of rapidity, the quickness of muscular exertion, and also through our sense of duration, including the element of fatigue. We can experience time both kinetically and statically, that is we can be conscious of moving quickly for a long time, or we can be conscious of pressing hard for a long time without moving: either kind of exertion can make us tired, and physiologists can explain tiredness in a chemical manner. We can also rest for a long time, and a certain duration is required for recuperation.

Time and Space therefore are not primary sensations, but abstractions deduced from the senses connected with exertion. Among these senses is the sense of work done, that is of the exertion of a force through a certain distance: another is the sense of power or activity: we know when we exert a force quickly, and can estimate the rate of doing work. We can run fast along a level, or we can push a truck slowly up-hill: the exertion may be the same; it will be estimated by our fatigue, using the term "fatigue," not in an exhaustive sense, but in a moderate and measurable sense. Any movement, however slight, if repeated often enough, will fatigue us. Hence an element of fatigue enters into any motion.

When we contemplate exertion, before we actually perform it, we have a sense of energy, and we know that we can expend that energy in performing the task: we can estimate our energy by the amount of work we are capable of doing. All these notions—force, speed, work, power, energy,—are connected with and derived from our bodily and chiefly our muscular sensations: they are, as it were, direct apprehensions, and with them we are equipped for beginning the study of Physics. That science utilises these primary ideas, and applies them in the first instance, no doubt rather anthropomorphic-ally, to all accessible objects.

Animals are presumed to have the same sensations; so that we can easily speak of force, exertion, power, energy, activity, in their case also. Similar results can be achieved by machinery or apparatus of various kinds: and although we know that no sensation is involved in their case, it is an easy generalization to apply the same terms to what they are doing. And when we deal, not with machinery in the ordinary sense, but with the behaviour of atoms of matter and other familiar processes, the same terms can be applied in those cases also. We know when a piece of matter is moving quickly, like a cricket ball or a bullet: we know when it is exerting force, like a stretched elastic or a bent spring: we know when it is doing work, like a clock-weight or a water-wheel: we know when it has a good deal of energy, like a hurricane or a fire or dynamite: we know when it is working at high power, like a locomotive or a speck of radium or the sun: we know also when energy is stored without working, as in the rotation of the earth or a lake at the top of a mountain or a coiled-up spring or an electrified body or a piece of coal. And thus we have come to define these things accurately, to apply them in all directions, and to build up the science of Physics.

Thus Modern Physics is limited to the behaviour of those modified portions of ether which we speak of under the generic name "Matter," including the electrical constituents of which matter is composed: and by aid of the primary sensations experienced in our own bodies, which are the most familiar pieces of matter we know of, we have made some progress in ascertaining the uniform system of laws which summarise the behaviour of that portion of the ether to which we seem to have direct access. Whether the access is as direct as we think it is, may remain to be doubted later on; but hitherto the science of Physics has been afflicted with no doubt of that kind. Nor, so far as I know, has Philosophy,—though on that of course I speak under correction.

THE ETHER AND ENERGY

Meanwhile we are sure of this, that matter is built up of electric charges, that electric charges are capable of locomotion, that intrinsically they are energetic, that they exert forces on each other through the intervening space, that they move with a certain rapidity, that they possess a measurable amount of energy, and that in getting rid of that energy, by transferring it to other mediums, they do work and thereby achieve some result, and that all the activity in the physical universe is the result of the transfer of energy and consequent work done.

POTENTIAL ENERGY

So long as energy is stored, no work is done, and no result achieved, except maybe in geometrical configuration or the maintenance of the *status quo*. A coiled-up spring has energy, but, unless it begins to uncoil, it does no work and achieves no result: gunpowder is passive and quiescent until it is exploded: a piece of coal does nothing until it is put on the fire. A planet sailing through space in a circular orbit is altering the geometrical configuration of the Solar System, but is doing no work and not altering anything else. Even a flying bullet, if you eliminate the resistance of the air, is doing nothing until it strikes a target. A spinning fly-wheel, if without friction, is achieving no object until it is harnessed by a belt over a pulley to some machinery.

Thus energy has two forms, a static form and a kinetic form, both illustrated in the examples chosen: in both cases energy is conserved. Work requires two factors, force and motion: either factor may exist without the other. Both may represent the possibility of doing work, and if so, they both constitute energy. One is static energy the other is kinetic: one is as real as the other. A force alone, however, without any power of continuance over a perceptible range, like a pillar supporting a roof, has no energy. The force must be able to continue when relaxation has begun. A strained elastic spring has energy, because the force endures: it does not fall immediately to zero. So it is also with motion. The motion of a point or of a geometrical figure or, by way of illustration, a feather or a ball of cotton wool, has no appreciable energy, because it can be stopped dead without exertion and without achieving anything. A railway train or a cannon ball cannot be thus stopped: it has not only motion but momentum: it possesses the mysterious property of inertia: its energy has to be got out of it by opposition, and in parting with its energy it may do damage. It achieves something: it parts with its energy to something else: it has done work. Kinetic energy is not motion alone, but motion with inertia, technically called momentum, that is motion with the power of continuance. Similarly, static energy is not force alone, but force plus the power of continuance, symbolised by a raised weight or a stretched elastic.

We shall find that kinetic energy resides, primarily at any rate, in matter, and therefore appeals more readily to our senses: it makes its appeal to our sense of motion. Static energy, on the other hand, because it makes a less direct appeal, as for instance in cordite or any kind of fuel or even a raised weight or a coiled-up spring, is sometimes called "potential energy," as if it were not actual and real: it is just as real as the other; and in many cases it appeals, or can be made to appeal, to our sense of force; but it resides, not in matter, but essentially in the ether; that is in the intervening space between the particles, in the substance which holds the particles together, the substance which is really strained and which can recover elastically from the strain. The energy of a coiled-up spring is not in the matter of the

THE ETHER AND ENERGY

spring, the particles have only been altered in relative position; all they have undergone is motion when the spring was wound up: the cohesive substance between them is what has been strained, and it is that which will recoil, it is there where the energy lies. So also with a raised weight: the energy is not in the weight, which has merely been moved, nor yet in the earth, which has been unaffected; it is in the gravitational field of force, which is able to push the two bodies together, that is, it is in the ether. All static energy is in the ether. All kinetic energy is in matter. That is a first rough statement, which may have to be modified somewhat in some exceptional cases; but eliminating those cases (the chief of which is radiation and possibly magnetism) matter alone is subject to locomotion: the ether alone is subject to strain. And the ether which is strained is not the modified ether, for that is matter: the ether which is strained is the main body of unmodified Ether, the substance which unites the stars and fills all space.

CHAPTER XI

FARADAY'S CONCEPTION OF THE ETHER

"Whatever difficulties we may have in forming a consistent idea of the constitution of the æther, there can be no doubt that the interplanetary and interstellar spaces are not empty, but are occupied by a material substance or body, which is certainly the largest, and probably the most uniform body of which we have any knowledge."

From article on "Ether," by CLERK MAXWELL, in *Encyclopædia Britannica* Ninth Edition.

Faraday's views on the Ether of Space were doubtless coloured by those of his predecessor Thomas Young, who himself owed much to Newton. Faraday however was so thoroughly immersed in phenomena that he could not help forming views of his own. He strongly emphasised the importance of the so-called "empty space" near and surrounding electrified and magnetised bodies. He may be said to have been the first in modern times clearly to perceive the important part which "vacuum" played in all phenomena, and to prove that a body which acted at a distance did so by modifying the medium all around it. The introduction of ponderable matter into the space only modifies, or it may be helps to demonstrate what was there before. The "temperature" of any part of space has to be defined as the temperature which would be exhibited by matter if placed there. In explaining diamagnetism, Faraday found it necessary to assume that certain forms of matter have what we should now call less permeability than empty space;

104

he likens the lines of force to rays of light, and he will not tolerate the idea that a magnet *in vacuo* is in a state of "magnetic darkness."

In many cases he was able to show that the introduction of matter into the space near a coil or a magnet made no appreciable difference to the magnetic field, unless the substance introduced was itself magnetic. His investigation on this subject is contained in Section 1,709-1,736, near the end of Volume I of his "Experimental Researches." In this early portion he limits himself to a record of observed facts, without speculating on their causes. Trained as a chemist, his natural instinct would be to assume that everything was dependent on material bodies; and only by repeated trials did he satisfy himself that matter *per se* often had no effect, and in any case was not essential, whether it was an insulator like shellac or sulphur, or a good conductor like copper. In other words the inductive influence, which he had himself discovered, was transmitted across empty space. The effect was however conceived by him as due to stresses and strains in the medium, which he mapped out by his lines of force. And hence he felt constrained, as we still do, to postulate an etheric medium of some kind, able to sustain those stresses and undergo those strains.

Incidentally I note that he contrives an arrangement exactly analogous in principle to Hughes's induction balance, and forcibly expresses his surprise at finding that a conductor like copper, in which induced currents must have been generated, still made no difference to the ultimate magnetic induction transmitted through it. He surmises however (truly) that there might be a momentary effect, which he had perforce overlooked owing to the sluggishness of his galvanometric detector. He states this clearly in Section 1,730. If he had had a telephone able to follow all the temporary fluctuations of current he would have arrived at Hughes's induction balance at this ancient date, 1838.

Led by experiments such as these, and by a multitude of others which tend to be forgotten now because they are recognised as perforce giving negative results (all which however had to be tried then, to be sure that he had overlooked nothing that his instruments were able to show), Faraday proceeded to speculate on a phenomenon *known* to be transmitted by empty space, namely the rays of light, and to ask what relation radiation had to his lines of force. He went on to consider the lines as existing not only in an electric and a magnetic field, but in a gravitational field also. He could not rid himself of the instinctive assuredness that there must be some connexion between all the various forces of Nature. One of these connexions he had the happiness of discovering, at a time when no theory indicated such a phenomenon. He found that a magnetic field exerted a rotational action on light sent along the lines of force. He tried to find an effect also when the field was applied to the source of light, but the instruments at his disposal were inadequate; the discovery was made by Zeeman long afterwards, when it was immediately understood by the mathematicians Larmor and Lorentz. Dr. John Kerr also, by beautifully conducted experiments, succeeded in capping the discoveries of Faraday by two more cross relations between electricity, magnetism, and light, which are known by his name.

The other still more minute phenomenon which ultimately brought gravitation also into relation with light, eluded Faraday altogether, though he earnestly sought for a gravitational effect of some kind on electric or magnetic or luminous phenomena. It eluded all other experimenters in our own day, and only succumbed at last to the mathematical genius of Professor Einstein.

FARADAY'S CONCEPTIONS

That Faraday gradually became thoroughly convinced of the existence of some continuous medium, intervening between material particles and transmitting the force, is made clear by reference to the stages through which he passed, first in electricity, then in magnetism. (See his early work recorded in Volume I of his "Experimental Researches," 1613–1616, and then again at 1663, 1710, and 2443.) But he has doubts whether his own lines of force would not serve the purpose of transmitting vibrations better than a uniform medium; in other words he perceives that the ether must have a structure, and not be like air or water. He expresses this in a letter to Phillips called "Thoughts on Ray Vibrations," in May, 1846. He not only finds lines of force useful, but insists on their real physical nature, and emphasises the fact that they exist round currents as well as between magnetic poles. He speaks in 3301 of Euler's view that

"The magnetic fluid or æther is supposed to move in streams through magnets, and also the space and substances around them."

He also speaks of lines of power within the magnet, and says they might be considered either as "currents, or progressive vibrations, or as stationary undulations, or as a state of tension." He quotes with approval Newton's doctrine about the impossibility or absurdity of action at a distance with no medium of communication, and says that "even gravity cannot be carried on, to produce a distant effect, except by some interposed agent fulfilling the conditions of a physical line of force."
One of his arguments is the intrinsic curvature of the lines of force, in that respect differing from rays of light. Another is based on the current induced by cutting them.

MY PHILOSOPHY.—PART II

"The mere fact of motion cannot have produced this current: there must have been a state or condition around the magnet and sustained by it, within the range of which the wire was placed; and this state shows the physical constitution of the lines of magnetic force."

But about the kind of substance or agent that will satisfy the conditions and cause the observed result, he has no idea and refrains from speculation. Indeed he is rather modern in the conclusion of a paper in June, 1852, and in the cautious way he speaks of space.

"Whether it [magnetic force] of necessity requires matter for its sustentation will depend upon what is understood by the term matter. If that is to be confined to ponderable or gravitating substances, then matter is not essential to the physical lines of magnetic force any more than to a ray of light or heat; but if in the assumption of an æther we admit it to be a species of matter, then the lines of force may depend upon some function of it. Experimentally mere space is magnetic; but then the idea of such mere space must include that of the æther, when one is talking on that belief; or if hereafter any other conception of the state or condition of space rise up, it must be admitted into the view of that, which just now in relation to experiment is called mere space. On the other hand it is, I think, an ascertained fact, that ponderable matter is not essential to the existence of physical lines of magnetic force."

This means, or may now be interpreted as meaning, that whereas experiments can be described and calculations conducted without mentioning the ether, yet contemplation of an impalpable connecting medium of some kind becomes necessary so soon as we begin to philosophise on the physical details of the process under consideration; since ponderable matter is quite inadequate to explain them.

CHAPTER XII

MODERN GIBES AT THE ETHER

It must be admitted that among a certain number of advanced mathematical physicists whose work has mainly lain in the twentieth century, the ether is regarded with suspicion, or even with contempt. And some of the opponents go so far as to say that the nineteenth century idea of the ether has failed to establish itself, and that in consequence the whole idea of the ether is under a cloud, and is only upheld by a few antiquated supporters, who, though they are willing to admit many modifications in the original nineteenth century notions of an ether, feel the need of a medium capable of performing the functions attributed to it. They therefore strive to extend the bounds of physics, and supply the continuum,—often regarded as an abstraction under the name space-time,—with physical properties which shall enable it to perform those functions which otherwise they cannot account for.

The supporters of an ether are not content with the appealing to it for the exercise of these functions: they endeavour to see whether it is not competent to account for the physical aspect of other phenomena running through all physics, and for other phenomena such as life and mind, which have not hitherto been taken into account by physics. In this effort the supporters of an ether are at some disadvantage, owing to the undeniable progress which has been made in the advancing physical science of the twentieth century without explicit reference to an ether. This advance has mainly been rendered possible by the free use of mathematical equations and other abstract methods of reasoning, which avoid the necessity of entering into detail concerning the mechanism. The Modern School of physicists do not seek clear ideas as to the processes involved. but are able to formulate them with sufficient vagueness to cover a multitude of methods of attaining observed results, and yet with sufficient definiteness to enable them to predict other results which may be anticipated, and which in some cases have been looked for and observed. This successful though difficult enterprise has lent twentieth century physics great éclat, and enabled its devotees to feel satisfied with their abstract methods, and to feel scornful of the old plan of trying to form a mental image of the details of what is happening. The old plan was not to feel satisfied with a mere abstraction, or with anything short of making or imagining what they used to call a "model" of the relevant activities.

For myself, though I am lost in admiration at the brilliant achievements of this Modern School, I cannot think that their philosophical outlook will be found ultimately satisfying. For the time being they are in the ascendant; and their methods are proving fruitful of results. But the abstract is a sort of compulsory process of tying down phenomena to ascertained laws, so that certain results are shown to be inevitable without any real understanding of how they are brought about. In the nineteenth century methods of this kind began to be used. Results were obtained or predicted by the law of conservation of energy, and by the second law of thermodynamics, which were powerful blindfold methods of arriving at phenomena subject to those laws. The predicted phenomena were afterwards verified as actually occurring, and therefore fully justified the process of arriving at them from the mathematical point of view. But they never gave a full and satisfactory explanation, such as is required to satisfy a philosophical physicist who seeks to understand the intricacies and inner workings of nature. To say, for instance, as Helmholtz did in 1847, that the induction of electric currents was an inevitable consequence of the conservation of energy, was not as illuminating as the more ambitious and really more difficult treatment of electro-magnetic induction in terms of the fields which were mapped out pictorially by Faraday's lines of force. It was thought no doubt by some physicists that the lines of force were supernumerary and superseded by a more abstract method. Supporters of the more concrete mode of treatment had no quarrel with the abstract method as far as it went; they fully admitted the inevitable deductions which it made possible, and yet they felt that there was more behind, and that exploration into the details of the action could be advantageously pursued. To this day I understand that a great physicist like Sir J. J. Thomson holds, as Faraday held, that the lines of force are not mere auxiliaries, like the lines of latitude and longitude on a map, but that they are physical realities.

And so I hold that the continuous medium, filling up the interstices between the discrete atoms of matter,— and indeed penetrating the interior of those atoms, and extending through all perceptible space,—will be found necessary too, before all the results treated of in twentieth century physics can be really and thoroughly understood. In this view I have the support of some of the twentieth century physicists themselves. I have elsewhere extracted utterances from Professor Einstein, showing that to some extent he shares this view. (See page 137.)

MY PHILOSOPHY.—PART II

The name "ether" for this great continuous medium is perhaps a little unfortunate, since it has been adopted by and is popular with a number of cranks, who find its vagueness rather helpful in their somewhat random speculations, and who tend to bring the term into contempt. The term itself is at least as old as Newton, who in his optical queries perceived the need for some such medium, not only in optics, but in connexion with gravitation too, in order to avoid the impossibility of action at a distance. It was adopted at the beginning of the nineteenth century by the brilliant exponents of the undulatory theory of light, in France and elsewhere: and throughout the century many attempts were made to explore its properties, on the whole without much success. Hence it is sometimes urged, on those who still uphold the notion, that some other name for it might avoid confusion. Another unfortunate and unnecessary confusion has been introduced into the subject by the carelessness of chemists who in order to designate a class of chemical compounds, which happened to be peculiarly limpid or to possess small viscosity, styled them ethers; so that they had a series of ethers, just as they had a series of alcohols, each of them derived from the radicals methyl, ethyl, propyl, butyl, amyl, etc.; a nomenclature of a trivial kind, which would hardly have been allowed by the physicists if they had been consulted; nor by the chemists if they had taken the ether of space seriously.

MULTIPLICITY OF ETHERS

One of the methods of showing contempt for the ether of space is to suggest that for the different phenomena to be explained, many ethers are needed. Sir James Jeans even goes so far as to suggest that the ether is a subjective notion, peculiar to each individual, so

MODERN GIBES AT THE ETHER

that every observer carries his own ether about with him; while another has suggested that an ether for Sirius would differ from an ether appropriate to the earth. All this I utterly deny, and pin my faith on one single ether, which must have properties sufficient to accomplish everything demanded of it. A multiplicity of ethers is of not the slightest use in physical explanation. To adopt a multitude would be to abandon any attempt at explanation, just as much as would be the abolition of all. A single medium is all that is necessary, by whatever name it is called. And if it is found insufficient to account for all phenomena, it must be given up. In that case it seems to me that all attempts at really understanding physics must be abandoned too, and we must rest satisfied with mathematical abstractions. I do not call the ether an abstraction; it is an inference from observed phenomena, just as matter is. There are those who hold that matter is a mental abstraction, along with space and time, and not a physical reality. Well, I am content to put the ether into the same category as matter, and if its denial only means that its reality is denied in the same sense as the reality of matter is denied, I cannot say that I am content with either denial, but I am willing to leave that thesis to the judgement of humanity.

APPEAL TO THE JURY BY THE COUNSEL FOR THE OPPOSITION

Among the great mathematical physicists of the present day I believe that Einstein and Eddington are more or less in agreement with my main contention. The chief opponent is a man for whom I have a profound respect, Sir James Jeans. He is an opponent worthy of attack, inasmuch as he has caught the ear of the public as well as of the scientific world; and he is vigorous in his contempt for the idea of an ether of space. He may be regarded as chief Counsel for the opposition, and his expanded Rede Lecture called "The Mysterious Universe" may be regarded as an address to the jury, in which he has summarised all the objections, and set them forth in an entrancing manner. Accordingly I feel called upon to appeal to the jury in the opposite direction, and reply to the Counsel for the Prosecution. I will refer to pages in the first edition of this book.[1]

On page 89 he writes as follows:

"Newton had realised that without an all-pervading ether, it would be impossible to determine the absolute speed of motion through space, and had also seen that such a medium would provide an unmoving standard by reference to which the motions of all bodies could be measured."

Precisely. That is what I hold to be exactly true, and that I hope may some day be determined. Since Newton's time, many attempts have been made to measure the speed of motion of the earth through the ether. Hitherto they have failed, and now, says Sir James Jeans, speaking of the ether,

"Einstein at one blow deprived it of its most important property of all, that of providing a standard of rest, by reference to which the true speed of any motion could be measured."

In other words, we have at present no criterion for absolute rest, and therefore none for absolute loco-motion.

But how did Einstein inflict this blow? Not by experiment, nor by calculation, but by a postulate.

"In 1905 Einstein propounded the supposed new law of nature in the form—'nature is such that it is impossible to

---

[1] It may be convenient to notice that pages 89 and 90 of the first edition become pages 79 and 80 in a still cheaper new one.

MODERN GIBES AT THE ETHER

determine absolute motion by any experiment whatever.' It was the first formulation of the principle of relativity." P. 89.

And later he says, on page 90:—

"Einstein's principle now tells us that, so far as all the observable phenomena of nature are concerned, we are free to define 'absolute rest' in any way we please."

And again, on page 91:—

"The principle of relativity assures us that all the phenomena of nature in this room are absolutely unaffected by this 1,000 miles-a-second wind, and would indeed be just the same if the wind blew at 100,000 miles a second—or indeed if there were no wind at all."

Well, granted the postulate, that would follow. But why should we appeal to a postulate as if it were a law of nature. The postulate says that a certain experiment is impossible. There are many cases where people have said that an experiment was impossible, and held to it until the experiment was actually performed. I for one am, and there must be many who still are, hopeful that absolute motion will one day be determined. It need not invalidate the results deduced from Einstein's principle, for undoubtedly in present-day physics the idea of absolute motion is absent. The phenomena are all independent of it. And accordingly it is a good principle on which to take a temporary and practical stand. All we know at present of locomotion is the locomotion of one piece of matter with reference to other pieces. That can be freely granted, and the whole edifice of physics is consistent with the idea that loco-motion through the ether is impossible to measure. But the impossibility has never been proved, and some day an exceptional phenomenon may be found. When it is found, the postulate will be given up, but the results deduced by its aid will still remain.

MY PHILOSOPHY.—PART II

The one thing we do know about the ether is that it is the transmitter of radiation, and that it transmits electromagnetic waves at a certain definite pace, a pace which Einstein holds to be the one absolute velocity in the universe. We also know that it is the medium responsible for electric and magnetic fields, and that radiation is due to the interaction of those two fields: moreover we are familiar with radiation not only from the possession of eyes, but by the possession also of what are called wireless sets. Strangely enough, however, Jeans seems opposed to this. After proposing some general arguments, he says, on page 94[1]:—

"There is, however, a stronger case than this against supposing the luminiferous ether to transmit radiation and electrical action."

The argument in favour of this strange thesis must be read in the book. It consists in pitting the law of gravitation against the laws of electromagnetism. He says, on page 95[1]:—

"An ether which transmits electrical action can hardly transmit gravitational action as well, since all the properties with which we can endow it are used up in accounting for its transmission of electric and magnetic forces."

And he goes on to suppose that the FitzGerald-Lorentz contraction would not apply to gravitational measurements, by arguments which I confess I do not follow. They seem to depend on the impossibility of gravitation being transmitted through the ether. Whereas, according to my view, if it could not do that, it could do nothing. The ether was originally invented to account for gravitation, only subsequently it was found necessary for light too, and, later on, for electricity and magnetism.

[1] Passages quoted from pp. 94, 95 will be found on page 83 of the new edition.

116

MODERN GIBES AT THE ETHER

The ether has to be competent to do all these things, and to be responsible for cohesion likewise. It is the one welding entity which unites all the particles of the universe, big and little, and converts them into a comprehensive cosmos.

There are other observations in this chapter, called "Relativity and the Ether," with which also I disagree. They relate to the similarity, and apparently essential identity, between time and space, a doctrine based on Minkowsky's well-known assertion—which after all is only an assertion—that

"'space and time separately have vanished into the merest shadows, and only a sort of combination of the two preserves any reality.'" P. 103[1].

Jeans supplements this by the following destructive sentence:— P. 103[1].

"This shews in a flash why the old luminiferous ether had inevitably to fade out of the picture—it claimed to fill 'all space,' and so to divide up the continuum objectively into time and space. And the laws of nature, not recognising such divisions as a possibility, cannot recognise the existence of the ether as a possibility." P. 103[1].

His contention is that time and space only differ subjectively. He says himself that the four-dimensional continuum is purely diagrammatic. Its division into space and time is not objective, it is merely subjective. This seems to mean, as Minkowsky presumably meant, that time can be treated as a fourth dimension of space. Now undoubtedly time and space can both be expressed in one equation, just as real and imaginary quantities can be similarly welded together. The variables are not only x_1, x_2, x_3, representing the usual x, y and z, or length, breadth, and height, but are

[1] Quotations from p. 103 occur on pp. 90 and 91 of the new edition.

117

supplemented by a time term called x_4, on a par with the others. But this x_4 represents time multiplied by a speed, and is kept separate from the others, by its containing implicitly, when finally you come to interpretation, $\sqrt{-1}$ as a factor, so that, by a customary mathematical device, two equations can be written as one. But that does not in the least mean that time and space are the same thing. To convert time into space it must be multiplied by a velocity; and if the only recognised velocity is c, it must be multiplied by c. The idea of speed is essential in any relation between space and time; in fact time is an inference deduced from speed. When you ordinarily say that a town is two hours away, you usually mean if you travel by a motor car. If you are referring to some house at a trivial distance, you would say "Two minutes' walk." Always the speed of the vehicle is involved. On that principle the sun is eight minutes away, Sirius is twenty years, the Andromeda nebula 800,000 years. But this involves no confusion between space and time: the light-year is a thoroughly understood simple unit. Eddington has shown that space and time can never be properly confused; they are essentially different. Only by a mathematical artifice can time be treated as a dimension of space.

Time may be a creation of thought, but not the ether. Jeans says of the ether:—

"Its existence is just as real, and just as unreal, as that of the equator, or the north pole, or the meridian of Greenwich." P. 105[1].

He implies that we can only think of it as a pure abstraction.

"It is at best 'a local habitation and a name.'" P. 105[1].

[1] 105 becomes pp. 92 and 93 in the new edition.

MODERN GIBES AT THE ETHER

But he goes on to say that the universe is being found to consist only of waves, and we naturally ask, Waves of what? He speaks of heat waves and suicide waves, as if they were analogous to waves propagated through the ether; just as a disease of potato-blight might travel across England as a wave from west to east. "Potato plants are not the medium of its passage," he goes on to say, "for they are not continuous." [No more are the particles of matter continuous.]

"They merely indicate the extent of its progress, like specks of dust in a ray of sunshine." P. 106.

I entirely agree that waves demand a continuous medium, and that matter is only an index of their arrival, and that "we have no knowledge of the waves except where there are particles of matter to reveal them to us." That is characteristic of the function of matter. It displays activity otherwise occurring. But I do not agree at all that nothing "is propagated that is more concrete than a mathematical abstraction," as noon may be said to travel over the earth.

Everyone holds that light really does come from the sun, and takes eight minutes on the journey. Jeans seems to wish to deny this, and has a dramatic argument between a physicist and a mathematician on this very point. The physicist is worsted in the argument, because he is gratuitously assumed to be in agreement with the mathematician, that

" 'at rest in the ether' means nothing at all, and 'a steady speed of 1,000 miles a second through the ether' means nothing at all." P. 108.

If this is a valid argument, it may be asked, Why introduce anything about what would happen to an observer moving at 1,000 miles a second, if that means nothing at all. And what difference can "nothing at all" make in the passage of light from sun to earth. There is little hope of understanding really complicated things, if things of the utmost simplicity have to be artifically complicated in order to uphold what may be regarded as an artificial and essentially irrational or unphilosophical scheme of nature. The "law of Minkowsky" is not yet a law of nature.

So there we return to the old crux whether absolute motion has any meaning. When that is ultimately determined, these questions will not arise. The theory of relativity will not be damaged, but it will be put in its proper place as a partial and temporary solution, of remarkable power, adapted to our present knowledge. At present Jeans seems to use it as a kind of bludgeon to coerce people to his views. He says, on page 111:— (page 97 in the new edition.)

"The three dimensions of space and one of time enter as absolutely equal partners into the formulation of every natural law. If they did not, the law would be at variance with the principle of relativity."

And he goes on a little later:—

"Space and time as separate entities have already disappeared from the universe; gravitational forces now disappear also, leaving nothing but a crumpled continuum."   P. 112.

I don't object to a crumpled or warped continuum as a mode of expressing gravitation. But how can a thing be warped or crumpled if it has no objective existence? He looks forward to a time when electromagnetic forces will also

"be resolved merely into a new type of crumpling of the continuum, essentially different in its geometry, but in no other respect, from that whose effects we describe as gravitation."   P. 113.

MODERN GIBES AT THE ETHER

"If so [he goes on] the universe will have resolved itself into an empty four-dimensional space, totally devoid of substance, and totally featureless except for the crumplings, some large and some small, some intense and some feeble, in the configuration of the space itself."   P. 113.

"The passage of sunlight from sun to earth, now reduces to nothing more than the continuity of a corrugated crumpling along a line in the continuum which extends over about eight minutes of time and about 92,500,000 miles of length. We now see that we cannot picture it as the propagation of anything concrete or objective through space unless we first divide the continuum objectively into space and time, and this is precisely what we are forbidden to do."   P. 113.

Not "forbidden" by the laws of nature, but by a conclave which has the power of putting things on the Index Expurgatorius, which must be appealed to, until it is willing to remove this embargo in the form of a Nihil Obstat.

CHAPTER XIII

THE PHYSICAL ASPECT OF THE UNIVERSE

An Alternative Scheme to that of
Sir James Jeans

In the January, 1932, number of the quarterly journal published by The British Institute of Philosophy, called *Philosophy*, Sir James Jeans with extraordinary ability represented the view of the universe which may be held now in the twentieth century by a mathematician, and concludes that this representation contributes to and upholds an idealistic philosophy. Now with the contention that an idealistic philosophy is superior to any other, that is to say nearer the truth, we may be allowed to sympathise. Several physicists, even in the nineteenth century, were inclined to sustain the essence of Berkeley's view, or to consider that metaphysical truth must lie somewhere in that direction. G. F. Fitz-Gerald, for instance, was notably of that opinion. But he was not prepared on that account to abandon the physical view of existence, and to take refuge in mere mathematical abstractions.

The mathematical method has a surprising power of making deductions about the result of any given activity; but when called upon to elaborate the actual details of the process, and construct a visible picture of how things interact, and trace in detail the course of their activities, it conspicuously fails. This failure is recognised by the mathematicians themselves. They know that their symbols can represent a number of different things, and that their equations can be correspondingly interpreted in various ways; but they do not find it necessary to interpret them at every stage of the process; nor need the mathematical transformation have any resemblance to the actuality of intermediate stages. Mathematicians can arrive at a result without forming a mental image or model of what is really happening; and so long as they attain a result which is intelligible, both at the beginning and at the end, they are satisfied. They have no need to follow the working; they go so far as to say that those who, in order to make a clear mental image of a physical process, postulate a mechanism not accessible to the senses and try to understand its actual working, are out of date and are left behind. Sir James Jeans seems to reject anything that the senses are incompetent to display, saying for instance:—

"The assumption that things existed which could not be perceived had led them into a whole morass of inconsistencies and impossibilities. The new policy was not adopted of set purpose or choice, but rather by a process of exhaustion. Those who did not adopt it were simply left behind, and the torch of knowledge was carried onward by those who did."

Yet it must be admitted that nearly all our progress in physical science up to the year 1900 was made by men who adopted this very machinery. It is one thing to say that mechanism is not a complete explanation, or that it is difficult to follow; it is quite another to sweep it all away and say that there is no mechanism at all, that we must give up the idea of making a vivid picture of what is happening, and must content ourselves with abstractions, or what are called "mental concepts." I should hesitate myself to call them mental concepts unless they can be conceived: and I do not perceive that any serious effort is made to conceive them. I should prefer to call them inconceivable or unconceived mathematical abstractions.

The old leaders were as well versed in mathematical methods as is the New School. The nineteenth century contained such men as Tait and Thomson, Stokes, Clerk Maxwell, Helmholtz, and many others. But they were not satisfied with the mere working out of their equations; they did not regard that as a complete statement of fact; they endeavoured to form a mental image or model of the working, and were thus led, not only to final results, but to some partial understanding or partial misunderstanding of the process by which they were achieved. In following this method they had, it is true, to postulate or infer the existence of entities which did not appeal to the senses, the laws of which therefore they could not explore experimentally, but which nevertheless they hoped to ascertain by further study as knowledge increased.

This utilisation of hypothetical agencies whose existence was inferred by means other than direct sense apprehension, was begun, or at any rate adopted, (mainly in tentative or interrogative form) by Newton, and was continued with various modifications down to the twentieth century. It was not considered a disadvantage, at any rate not a serious one, that these things could not be actually perceived in themselves; they were perceived by their material activities, by the way in which their supposed activity operated on a material body. It was not considered necessary to put them out of court or deny their existence, merely because, like the corpuscular particles of light, they were directly inapprehensible. The whole doctrine of lines of force in space was of this nature, and formed the basis of the physical philosophy of Faraday and Clerk Maxwell. One effect of the lines could be displayed, in the case of magnetism, by iron filings; but the cutting of physical lines of force, as a means of expounding quantitatively the induction of currents, was a device only justified by results. The idea of a gravitational or an electrical field had no other justification. But that was not considered as a fact hostile to the existence of such fields; for there are many other things active in the universe which make no direct appeal to the senses.

Experience shows that some material bodies are animated, but the animation is only inferred from the behaviour of the material body itself. The nature of life is unknown; we only experiment on animated organisms, that is on the material machinery worked by life; but few of us are willing on that account to deny that life exists. There is indeed a School of biologists, who would seem to be approved by Sir James Jeans, who agree with him that the only proper object of study is the *behaviour* of organisms as displayed to us through our senses, and that all else is wasteful or meaningless hypothesis.

Twentieth century science, according to Jeans, having become aware of a space-time continuum—though how it became aware of such an abstraction by means other than hypothetical interpretation and inference is not obvious—shirks an attempt to draw a concrete picture of this continuum; but "under the guidance of Poincaré, Einstein, and Heisenberg," apparently came to recognise that in science the

"only proper objects of study were the sensations that the objects of the external universe produced on our senses."

Rather an amazing and solipsistic statement, this, if pressed: it might be true for psychology. I fancy it may be a slip of the pen,—if so I do not press it,—for we surely try to study the things responsible for our sensations rather than the sensations themselves. How

else can we gain a knowledge of an external world?
But then, Jeans goes on,—

"The dictum *esse est percipi* was adopted whole-heartedly
from philosophy—not because scientists had any predilections
for an idealist philosophy, but because the assumption that
things existed which could not be perceived had led them into
a whole morass of inconsistencies and impossibilities."

Well, that is a strong statement: especially if it be taken
to imply that things not perceived, or at least not per-
ceivable, do not exist. It would have very much surprised Professor Tait, not to mention Newton himself.
Professor Tait, the centenary of whose birth we might
have celebrated in 1931 along with the other centenaries,
did not hesitate to postulate an "unseen universe," that
is a whole collection of existences which make no direct
sensory appeal. But twentieth century science apparently sweeps these away, one and all,

"not from choice, but from necessity. They had to be swept
away, because their presence introduced confusion and incon-
sistency into the scientific picture of the world."

But it may be contended that the sweeping away is not
improving matters: we are left with no picture at all!
Denial of everything that cannot be directly perceived
by the senses seems to me unworthy of modern science:
we cannot really be satisfied with mathematical abstractions alone, even if we call them mental concepts;
nor need we limit the attention of science to things that
perturb the senses, still less to the sensations themselves!

THREE STAGES IN SCIENTIFIC HISTORY

The history of science is divided by Sir James Jeans
into three periods, of very different lengths, which he
calls respectively the animistic, the mechanical, and the

ALTERNATIVE ASPECT OF THE UNIVERSE

mathematical. The first period lasted from the earliest ages down to Galileo and Newton: the second period from Galileo and Newton down to A.D. 1901: the third period is the portion of the twentieth century we have so far lived through.

Well, when we consider animated matter, I am not at all sure that we shall not have to find some truth even in the doctrines of the first period. In fact I might put it more strongly: for my attention has just been called to the Gifford Lectures of Professor Stout called "Mind and Matter," in which he deals, on the whole favourably, with animistic views. However this may be, the second or materialistic period taught that

"inanimate nature appeared to behave as though its constituent pieces exerted pushes and pulls on one another, exactly similar to those we exert on them by the action of our muscles— it was in this way that the science of mechanics had its origin. Each piece of matter was supposed to exert a 'force' on every other piece. . . . The force which A exerted on B was equal in amount to that which B exerted on A. When the forces were gravitational or electrical, the bodies need not be in contact."

(I should not express the facts precisely in this way, and will deal with the discrepancies directly.)

He then goes on to say that as knowledge increased, or

"as science began to wander farther from home, its inferences as to the nature of the assumed machinery became first confused, and then contradictory and absurd, until finally there was only one inference possible, namely, that the assumed machinery did not exist at all." The idea that objects pushed or pulled each other about "was as much an anthropomorphic error as the earlier animistic universe of our primitive ancestors,"

for in it the analogy of whims and caprices was replaced by the analogy of muscles and sinews. So, to evade the error, instead of searching for the mechanism elsewhere, the idea of "force" was abolished; and, still more

127

drastically, "twentieth-century science, penetrating to the farthest depths of the universe, has swept away" every part of the machinery.

Now I know that this is the twentieth century point of view, and I too hold that it is erroneous to contend that one piece of matter pushes or pulls another about; I know that there are many difficulties about the exertion of force by one piece of matter on another; at least so long as they are not animated, as in a football scrimmage. It is contrary to what I have been just emphasising, that inanimate matter is absolutely inert, and devoid of all activity. But there may be a way out, other than the mere abolition of what after all are manifest activities. Force, or rather stress, is a reality directly apprehended by our muscular sense. We experience it in a nut-cracker or a thumbscrew or a rack: we see it in the oscillations of a falling drop. The idea of force cannot be really replaced by acceleration, its effects are different, save when it it quite unimpeded or when the body acted on is small or rigid: and yet it is true that a material body A does not exert direct force on another material body B. Material bodies never really act on each other, for they never touch, and they cannot act at a distance. Newton knew they did not, and could not, when he promulgated the law of gravitation. What he really meant was that their motions could be investigated on the assumption that they were the same as if they did so act, particle by particle. The law of gravitation is a philosophy of "as if:" so is Coulomb's Law.

In Jeans's statement quoted above (on page 127) about pieces of matter and the force they exert on each other, I accept his statement about what used to be said concerning exertion of force, but I wish to deny that directly acting bodies are "pieces of matter." I also deny the concluding sentence that when the forces are gravitational or electrical the acting bodies need not be in contact. I say, on the contrary, that things that exert force on each other directly must always be in contact, and that any force observed between two separate pieces of matter must be transmitted by something which is in contact with both. When a gun fires a shell, the gun has no direct action on the shell, although the momenta are equal and opposite: the cordite fires both in opposite directions. And its mass must be taken into consideration for accurate specification, unless it is either zero or infinite. There is always something between two pieces of matter, whether the gap be perceptible or not. When something called A really acts on something called B, one of them may be a piece of matter, but not both. One of them may be the ether. If motion results, work is done, and the energy lost by A is gained by B. The energy lost by A may be kinetic: if so, the energy gained by B is potential, and represents a strain in the ether. On the other hand, when the converse operation takes place, the energy gained is kinetic. The energy is merely transferred, somewhat as in a commercial transaction "Capital" changes ownership, without loss or gain; until the act of purchase (or of work) is complete.[1]

There is a constant interchange of energy between ether and matter: what one loses, the other gains. One of the acting bodies is always the ether. All potential energy is therein stored, in imperceptible fashion, until it can be made manifest again by being transferred in kinetic form to matter. The action may be instantaneous (i.e. transmitted with the velocity of light), but the ether is an inevitable intermediary in every action between

[1] Economists can decide whether the purchase was justified: a transfer may be gratuitous, without adequate service rendered to anybody, as in gambling. So it can be also in physics; water can fall without achieving anything useful, and then there is waste, or increase of entropy. Energy which was tractable and therefore available, when possessed (as we call it) by high level water, has been degraded into the unorganised and uncontrollable molecular motions of low-temperature heat.

pieces of matter, whether it be perceptibly instantaneous, or whether it is obviously separated by space and time, as in a gravitational field. The existence of energy is continuous, and is just as real in the potential form as in the kinetic: only as it belongs to an entity which makes no appeal to the senses, its existency is not all the time conspicuous.

Unless the ether is mentally recognised, the idea of potential energy gives or ought to give some trouble. It has been sometimes called "possible energy," in contrast to actual energy, a nomenclature which really plays havoc with the law of conservation. Certainly the energy is in full existence all the time, and I believe will be found to have identity, the same kind of identity as is possessed by a piece of lost luggage; though it may not be so easy to recognise when it returns to our ken, nor is it so ready to be labelled. Yet everyone knows that our fires are liberating energy emitted by the sun of the carboniferous epoch, which, having arrived through the ether and been stored by vegetation, has been locked up quiescent for a vast number of millennia, and now at length is being restored to the ether as low-temperature radiation.[1]

Returning to the law of gravitation, and contemplating the motion of a planet or a projectile "attracted" by a central body, we know now, even if we didn't know before, that these "attractions" are due to something between or surrounding the bodies, something not sensed nor yet adequately conceived, which is warped by the neighbourhood of a large mass; or as Jeans says, the framework is twisted up, so that the path taken by a freely moving body becomes curved. "The framework" is his name for the space-time continuum in which

---

[1] A little further elaboration may be necessary for cases of vibration. See Lodge's article in *The Philosophical Magazine* for October, 1879, and June, 1881.

ALTERNATIVE ASPECT OF THE UNIVERSE

everything occurs. The influence of a warp or curvature in space is felt by a material body as a force; and if the body is large there may be some deformation. It appears to be a true doctrine that matter is inert and passive, never exerts force, or does anything else. Everything that happens is done to it. A leaf in the wind is a typical example. The wind itself is only displayed by its action on some form of matter: no one supposes that the kitten-like frolic of a dead leaf is inherent in itself. A piece of matter is perfectly inert, it always takes the path to which it is constrained—the path of least resistance.

> "The Ball no question makes of Ayes and Noes,
> But right or left as strikes the player goes."

It never chooses its own path, whether of least action, or any other. How can a particle choose its path to a given destination? Can we suppose it guided by the future? How can light choose the point of impact on a mirror, so that its selected path shall be the shortest possible? The principle of Least Action is not an explanation or guiding principle, it is no law of guidance, it is a statement of fact; it gives no reason, it is a blind-fold and interesting method of attaining a result, as the Conservation of Energy was in Helmholtz's brilliant paper of 1847. Applying the law of conservation, the intermediate steps of the process are ignored and unknown; a result is deduced, and is simply inevitable. That is the method of energy. It is a capital method in physics, but not one conducive to philosophical insight. Every idea of choice or self-determination applied to inanimate matter is obviously fanciful and figurative, though such language is often used in exposition. A particle cannot know, on setting out, which path is going to be the shortest or quickest or easiest. It cannot know what destination to aim at, nor by what route to travel. An inert thing must be guided instant by instant.

The Wave Theory has long shown how the route of radiation is determined, and by what means it arrives at a focus. Every wave must arrive in the same phase, if they are to reinforce each other. Every bit of a wave-front thus builds up the next wave-front. The time of journey may be a maximum or a minimum. All that is necessary is that every portion shall take the same time, else they will arrive in discordant phases. Usually that condition is satisfied by a minimum; and so the principle of Least Action can be used to determine the path. So the path can be predicted and it may look as if the thing moving knew what it was about. But note that the wave theory is not limited to ascertaining or determining the final result: it enables us to follow the process throughout. It gives a reason why the quickest path is the one taken. What is essential is that every part of a wave shall arrive in the same phase: if not there will be interference. Radiation is guided by the very properties of the medium. No question of choice or foresight arises.

This mode of accounting for "least action" is admitted to be true for radiation. How comes it that particles behave similarly? What determines *their* route? Well, that is really a discovery of twentieth century physics. Every particle is found to be accompanied by waves and is guided by them. Guidance in both cases is due to something etheric. The difference between a wave and a particle is getting partially obliterated. We have not yet heard the last of that curious discovery. It means that the behaviour even of a particle of matter cannot be understood without taking the ether into account. It may turn out that the very existence of matter will have to be explained in terms of the ether. The tables are sometimes turned in that way.

ALTERNATIVE ASPECT OF THE UNIVERSE

Modern Physics is very dogmatic, and it has much to be proud of, but it does not see all the way yet. Its defect is that it is too ready to be satisfied with foreseeing a result, and does not care to follow the details of a process, even though the process is essential to the attainment of a given result. How can a particle foresee and choose the path of least action? No answer can be given unless waves are introduced; and waves imply something that undulates; even if we choose to call that something "a nominative case."

TRUTH OF MECHANISM

In contending against the self-sufficiency of mechanism, our enthusiastic leaders are too ready to sweep away all machinery, and to take refuge or to rejoice in bare formulae and abstractions. Yet machinery there must be. Results are not obtained by rubbing a lamp or a ring, but by mechanism of some kind. This is true even for Creation; not only is mechanism necessary for attaining a result, duration is required too. The processes of evolution are slow and gradual; the element of Time is essential. Space will not do instead: Time and Space are essentially different. The ratio between them is not numerical but kinematic: the ratio is a certain fundamental etheric velocity. Our study of machinery is not ultimate, it is not self-designed or self-acting, but the machinery is *there*, and no good purpose will be served by ignoring it or sweeping it away.

SUMMARY

To summarise this portion:—

There is no real activity in matter: all the activity lies in what it is the fashion now to call space-time, or what I prefer to call the ether; which therefore must have some structure, though the structure is not yet ascertained. The only force exerted upon a body is that which the ether exerts upon it. The ratio of resultant force to mass is acceleration. There is a universal contact action, with equal reaction: the action and reaction are not really between material bodies, but between one body and the ether in contact with it. In the case of gravitation, every particle of a large body is acted upon. When the force is greater per unit mass on some nearer particles than on others more distant, the body experiences tidal deformation.

That in general has been my philosophy since the 'seventies of last century. And it is at least an alternative to the sweeping away policy adopted in the twentieth century. The great mathematical physicists of the twentieth century didn't understand how the mechanism worked,—didn't understand any better than those of the nineteenth,—but they discovered that they could get results without attending to any mechanism, so they proposed to dispense with it altogether, and apply the blindfold but powerful method of mathematics. For inanimate matter this plan works, though it gives us no picture of the happenings; but for animated matter it fails. Every biologist and physiologist knows that an organism is full of complicated machinery: it is the action of the mechanism that they are studying, and usually they have no short cut to determine what the result will be. They are not tempted to sweep away the mechanism, as a mathematician is: their temptation is to sweep away the animation, to deny that life exists apart from matter. Their philosophy is apt to be materialistic. They really do try to study the objects of the external universe wholly through the senses, and to ignore anything that cannot be thus perceived. Many of them have found however that pursuing this far enough leads to difficulties, and is not conducive to clear conception or explanation. The self-will of an animal is puzzling: they can seldom predict what it is going to do. They try not to admit original or "purposive" action of any kind, yet purpose is forced upon them by observed facts. Both modern physicists and biologists therefore are liable to make the same kind of mistake; they are liable to attend exclusively to the organism and ignore the thing in which the activity really resides. They are thus led to conclusions opposite in character but equally gratuitous. The biologist tries to think of mind as a function of matter, and to explain everything in terms of mechanism. The modern physicist ignores the mechanism, and having learnt that by mathematical reasoning he can attain results in the inorganic world without it, concludes that the universe is governed by mind. In that conclusion he may be correct. I say nothing against the idealistic philosophy. I sympathise with it. But I don't want to prop it up by false buttresses: it should be strong enough to stand on its own foundations.

THE POLICY OF THE CLEAN SWEEP

Yet I would not have it supposed that I approve the motto *esse est percipi*, with the implied correlative *non percipi non esse*, as if everything that could not be either actually or potentially perceived was scientifically to be considered non-existent.[1] The policy of the clean sweep is very drastic, somewhat fashionable, but seldom justified. It is the policy of the builder who cuts down trees before erecting cottages: it simplifies his task, but it spoils the village. There are alternative policies. Mechanism need not be swept away, but by perceiving that the ether is the most essential part of it, it may

---

[1] That "the sycamore tree continues to be when there's no one about in the quad" must be true both for science and for metaphysics: hence I insert the word "potentially."

gradually be understood. The properties and functions of the ether have still to be investigated. By this means physics and biology, which have drifted so far apart, can some day be brought together. The psychologist too can enter into his heritage; and the universe can be recognised as a complete whole, harmonious with the mind of man, who will not only anticipate part of its future course, but will be able to watch with a completer understanding every detail of the process. That is the sort of thing that can be attained by a really comprehensive knowledge: "the very hairs of your head are all numbered."

THE EVASION OF AN ETHER

With all deference to Sir James Jeans I want to contend that the Theory of Relativity does not abolish the ether. It only affects some of its imagined properties. An ether is still necessary, though it is not the ether of the nineteenth century: its constitution and properties have still to be made out. Apparently it cannot be moved, in the sense of locomotion: that is a function of matter. No one expects to "move" space: nor is what fills space subject to motion; it must behave as if it had infinite inertia. Ether is the seat of potential energy; it is the recipient of all strain; it alone exerts stress; it is the vehicle of an electric and a magnetic field as well as of all radiation. But in itself it is not like any other substance; nor is its behaviour explicable mechanically; (though whether some of it can be explained hydrodynamically, remains to be seen). The motions, the accelerations, of inert matter are not intelligible without it. It is essential to all activity. I rebel against much of the philosophy of Mach, and often against the philosophy of Poincaré. I take refuge in the brilliant philosophy of Einstein, to whom, in conjunction with Professor Max Planck, we owe such generalisations

ALTERNATIVE ASPECT OF THE UNIVERSE

as are suggested by $W=hv$. Let me quote a longish passage from Professor Einstein's lecture on Ether and Relativity given in the University of Leyden in May, 1920, and translated by Drs. Jeffery and Perrett as part of a book (published by Methuen) entitled *Sidelights on Relativity*:—

"There may be supposed to be extended physical objects to which the idea of motion cannot be applied. They may not be thought of as consisting of particles which allow themselves to be separately tracked through time. . . . The special theory of relativity forbids us to assume the ether to consist of particles observable through time, but the hypothesis of ether in itself is not in conflict with the special theory of relativity. Only we must be on our guard against ascribing a state of motion to the ether.

"Certainly, from the standpoint of the special theory of relativity, the ether hypothesis appears at first to be an empty hypothesis . . . but on the other hand there is a weighty argument to be adduced in favour of the ether hypothesis. To deny the ether is ultimately to assume that empty space has no physical qualities whatever. The fundamental facts of mechanics do not harmonize with this view. . . . Newton might no less well have called his absolute space 'Ether'; what is essential is merely that besides observable objects, another thing, which is not perceptible, must be looked upon as real, to enable acceleration or rotation to be looked upon as something real. . . .

"Therewith the conception of the ether has again acquired an intelligible content, although this content differs widely from that of the ether of the mechanical undulatory theory of light. The ether of the general theory of relativity is a medium which is itself devoid of *all* mechanical and kinematical qualities, but helps to determine mechanical (and electromagnetic) events. . . .

"As to the part which the new ether is to play in the physics of the future we are not yet clear . . . we do not know whether it has an essential share in the structure of the electrical elementary particles constituting matter. . . .

"Recapitulating, we may say that according to the general theory of relativity space is endowed with physical qualities; in this sense, therefore, there exists an ether. According to the general theory of relativity space without ether is unthinkable."

Surely that is a striking statement, coming from Professor Einstein, who cannot be considered antiquated and out of date.

THE POSSIBILITY OF OBSERVING ABSOLUTE MOTION

One reason why modern physicists are unwilling to accept the ether or pay any attention to it, is because they seem to hold the doctrine that the only things worthy of attention are those which make appeal to the senses and can be experimented upon. It is commonly said that all experiments on the ether yield negative results, by which is meant that they fail to give any answer, except zero. Special stress is laid upon the Michelson-Morley experiment, in which the question was put to nature, at what rate is the earth moving through the ether? the test applied being to compare the time taken by the to-and-fro journey of a light-beam in different directions. This comparison, it was thought, was bound to give a very small positive answer, of the second order of magnitude. And when this answer came out unmistakably zero, all our ideas were perturbed and had to be overhauled. A compensating cause was suggested by G. F. FitzGerald, depending on the way that matter might behave if electrically constituted and in motion. But that this compensation should be exact, as Lorentz on electrical principles showed it would be, was considered too remarkable to be satisfactory: and there were talks of a conspiracy among the forces of nature to prevent absolute motion from being discovered. So people were glad to take refuge in the Theory of Relativity, which *postulated* that such a motion was meaningless, or in other words, that the experiment

138

must fail,—as all similar experiments made directly on the ether had likewise failed.

But now I want to urge that perhaps we were not experimenting on the ether in the right way. We were aiming too directly at dealing with it. The ether does not appear to be a thing that we can directly deal with, it eludes our senses. There seems to be no way in which we can directly get hold of it. We have inferred its existence from the behaviour of matter, and that appears to be the right way to deal with the ether. Suppose we were trying to discover, by direct attack, any of the things that elude our senses, a gravitational or a magnetic field for instance, using ether waves as a test. Faraday's pertinacity was rewarded when dealing with a magnetic field, but the discovery was only made with the aid of a material body. The heavy glass in his experiment exercised a rotary action on a beam of light travelling along the magnetic lines of force inside the glass. With some substances the rotation was clockwise, in others it was counter-clockwise; so the effect observed was really an effect on a modification of ether inside matter, and was not made on free ether at all. In free ether the experiment fails.

Faraday also made experiments on a gravitational field, but these all failed. If he had been a modern physicist he might have said, "Sweep away the gravitational field: no such thing effectively exists." And yet the influence of a gravitational field on matter is conspicuous. The behaviour of matter near the earth displays the existence of such a field in almost every movement, though the effort is now made to express these movements in geometrical terms, without appealing to force at all. I should prefer to say that the neighbourhood of a large mass of matter so modifies the ether that the easiest path for a particle to take is the curved path that is actually observed. "Abolition of force" is now fashionable, since it seems unnecessary to explain the behaviour of matter, which in no case exerts any rebellion against it, and therefore does not need it to explain a curved path. This is a fallacy to which people are liable who are too familiar with a phenomenon, and led Professor Haeckel to regard the law of Conservation as obvious, not needing argument or evidence, and has led Professor Eddington to speak of the law of gravitation as "a put-up job."[1] The tendency is to forget that force is involved in every deflection of a thing possessing inertia, wherever it occurs. Motion along a groove or channel is easy and simple, but the law of force is not evaded. A warp in space does not curve the path of a body without exerting force, and therefore it must sustain the corresponding reaction. And recently, through the genius of Einstein, it has been discovered that light also takes a curved path when passing through a very strong gravitational field. Thereby the gravitational potential has for the first time been directly observed; but extreme conditions were required before it could be displayed: no ordinary experimenting would have shown it. The entire mass of the sun had to be pressed into the service; and the light had to come from a very great distance. By exceptional experimenting such as that, the ether itself may possibly be constrained to yield an observable effect. We have failed so far, perhaps only because we have not pushed the experiments far enough. We have perhaps been hasty in postulating that such an experiment will always be impossible. There may be a slow circulation of ether in an exceptionally intense magnetic field. It could be looked for, as affecting the velocity of light slightly in free space. It would be a delicate and difficult experiment; and the result is uncertain, as in any real

[1] Eddington, *The Nature of the Physical World*, p. 143.

experiment it ought to be. It *might* require circularly polarised light.

Meanwhile, as in the case of gravitation, the most ordinary properties of matter display the action of the ether when rightly interpreted. Not only is it responsible for the curvature of the path of a projectile, and the weight of a block: we observe its powers also when we experience the strength of a rope, the stiffness of a rod, the elasticity of a spring. In all these and many other cases we are really face to face with the properties of ether. Cohesion is just as much due to etheric influence as gravitation is; molecules are held together by etheric links, and when the links are stretched, there is strain; but the strain is not in the matter. Matter itself is never strained, it is only moved; some portion of it is relatively altered in location, the strain thereby caused is in the connecting immaterial mechanism between the particles. Therein lies both the elasticity and the tenacity. The atoms of matter are separate individual particles, only welded together and held in position by etheric forces, either electric or magnetic or both. Whenever we wind up a clock we are really experimenting on the ether, whether the clock be driven by a weight or by a spring. We feel the ether quiver when we are toasting ourselves at a fire or basking in the sun. Less obviously, when a gas condenses to a liquid, it does so by reason of the cohesive forces between the molecules; and again the ether is responsible. Hence a shower of rain is an indirect demonstration of an activity of the ether. Without the ether there could be no liquefaction, and the material universe would be a heap of dust, or rather a peculiar kind of perfect gas. So when people say that all experiments in the ether fail, that it makes no difference to us, and that it can therefore be wiped out of existence,—they only mean that we have failed so far in tackling it directly; whereas indirectly, through the properties of matter, that is by the interaction of material particles with it, its influence is among our most commonplace experiences.

THE ENGINEER'S VIEW

An engineer does not recognise this point of view; and the reason is plain. For him a girder or a tie-rod is not an assemblage of particles held together by a uniting mechanism, he does not thus analyse a block of matter, he treats it as a whole. He need not trouble to account for the fact of cohesion, nor need he ask himself what strain among the molecules is really like. There before him is a material body, with measurable elasticity, density, tenacity, and all its other properties. He need not account for those properties, he can just measure them. He used not to worry about crystalline arrangement, until he found that its gradual genesis weakened his structures; he was content with a practical issue. Thus set free, an engineer could build bridges, cut through isthmuses, regulate the flow of rivers, construct magnificent engines, and do other wonderful work. So it may be long before he has to attend to the ether; unless indeed his business is telegraphy of some kind. But a physicist, especially a philosophical physicist, has to attend to the minutiæ of atomic occurrences, and cannot be satisfied with the broad view rightly taken by the civil or mechanical engineer.

CONCLUSION CONCERNING THE INTERACTION OF ETHER WITH MATTER

The behaviour of mechanism cannot be thoroughly understood in terms of matter alone. One particle of

ALTERNATIVE ASPECT OF THE UNIVERSE

matter does not act directly on another. To have action you must have contact, and particles of matter are never in contact. If they ever get into contact, they cease to exist as matter, and turn into radiation. Their energy takes another form; from which apparently it finds a difficulty in escaping. Yet radiation can separate electrons from protons: a quantum of light can make an electron jump; and it may do other things, some of them perhaps not yet discovered.

Consider also ordinary locomotion. Locomotion of matter through ether is difficult to observe, even when it is moderately fast. Already motion is known to increase the mass of a moving body. Mass is admittedly a function of speed. How is that to be explained without taking the ether into account? What kind of locomotion can increase the mass of a body save absolute motion through space? and what sort of space can affect mass save one that has physicial properties? Moreover, when we combine two relative motions we find that to do so accurately, an ether-constant c must be introduced into the expression for the resultant speed. How can we account for that if we ignore the ether? We can express it blindfold in a mathematical formula, but we cannot give a physical interpretation unless we penetrate into the mechanism more deeply. If we attempt to explore the mechanism we may be led to further discoveries. On the other hand, if we ignore the mechanism, we are left stranded in the air, with no foothold in reality, nothing but what is called "a mental concept;" a strange term for something that we are unable to conceive.

Motion through the ether may be directly demonstrated some day, if we persist in looking for it. If we throw up the sponge prematurely and say that the whole idea is meaningless, we shall fail. When Faraday

143

failed to find the effect of gravitational potential on light or electricity, he did not throw up the sponge and say it was impossible. He urged that further experiment should be made. Ultimately a minute effect on light was discovered; indeed more than one; but only by great skill, elaborate apparatus, and by aid of the vast masses dealt with in astronomy.

CHAPTER XIV

VIEWS OF THOMAS YOUNG, NEWTON AND FRESNEL

That ether is a very substantial entity, far denser than any form of matter, has been gradually becoming clear to physicists. At first we only said that it must be denser than lead or gold or platinum, but we find that it must be out of all proportion denser. I have made an estimate of its density, in the light of electromagnetic theory, and it comes out inevitably huge. Every cubic millimetre contains as much substance as what, if it were matter, we should call a thousand tons. As the ether is not matter in the ordinary sense of the term, our ordinary units of measurement are inappropriate; but on the analogy of matter, the ether is of the order a million million times as dense as water. All its properties are of supernormal magnitude. Its rate of vibration, which enables us to see any ordinary object, is five hundred million million per second: a number so great that to try to conceive such a number of vibrations per second simply dizzies us. The number of seconds which have passed since ancient geological periods of twenty million years ago is about this number. Yet we familiarly make use of these vibrations. Our wonderful organ the eye is constructed so as to cope with them, in the easiest possible manner. And most people are as ignorant as are the animals of the strange etherial environment amid which we all live, the vibrations of which convey to us much information and awaken so keenly our sense of beauty.

To expound in ordinary language the way in which ether density is calculated is an attempt hardly worth making, but a general notion of its high value can be given by the reminder that atoms are extremely porous bodies and that the electronic specks which construct them are relatively to their size, as few and far between as the planets in a solar system. The average density of matter in a solar system is very small, for if sun and planets were spread out uniformly through a spherical volume of the same linear dimensions, they would constitute an extremely attenuated gas. So it turns out that any kind of matter is an extremely rarefied substance as compared with the particles of modified ether which are its real ingredients. An electron or proton must be millions of times denser than is the atomic system which they and their motions constitute. There is no absolute standard of density—everything is relative—and what we call the density of ordinary matter is almost infinitesimal compared with that of the continuous incompressible substance of which it is presumably composed but of which our senses give us no direct information.

The probability that matter might be extraordinarily porous was astonishingly recognised by the great genius of Thomas Young, who, writing in 1807, expressed himself thus:—

"The diameter of each atom [or electron as we should now say in this connexion] must be less than the hundred and forty thousandth part of its distance from the neighbouring particles; so that the whole space occupied by the substance must be as little filled as the whole of England would be if filled by a hundred men, placed at the distance of about thirty miles from each other. This astonishing degree of porosity is not indeed absolutely inadmissible, and there are many reasons for believing the statement to agree in some measure with the actual constitution of material substances."

OLDER VIEWS

The consequent great relative density of any continuous intervening substance was recognised by Newton; for in the fourth edition of his *Opticks*, he speaks of the Æthereal medium as a "fluid which fills all space adequately without leaving any pores, and by consequence is much denser than quicksilver or gold." It is true however that he goes on to argue (mistakenly) that this great density would obstruct the motion of the planets and hence he uses his perception of its great density as an argument against the existence of such a dense continuous substance. For this reason he reverts to his corpuscular theory of light, in order to evade what he foresees must be the consequences of a fully developed wave theory.

Until instructed, we can hardly help thinking of matter as dense and of ether as tenuous, but that is a poetic illusion associated with the term (sic) "ethereal." It is an illusion based on the testimony of our senses, which, as so often happens, has to be corrected by deeper insight into the real nature of things. Matter appeals to us so strongly, not because it is anything but a gossamer-like or milky-way existence in the vast continuity of ether, but because our obvious bodies are made of matter and because our animal sense organs are specially adapted to existence in association with matter, and give us information about nothing else. Even light, which we *know* is an ether vibration, tells us nothing about itself without study, what it tells us familiarly is—not about light, but—about the material objects which have emitted or scattered or differentially absorbed it. We get this information by life-long, indeed age-long, inherited and instinctive experience; we interpret the luminous indications without difficulty, and we forget the strangely complex nature of the processes which underlie all our channels of information; we only find their true nature out when phenomena are fundamentally analysed and seriously cross-questioned. When we have pursued this line of investigation for many years, we find that the important thing in the physical universe is ether, and that matter is trivial in comparison. Yet we can freely admit that matter takes such splendid and beautiful forms that it is worthy of the continued study of generations of scientific men; and we need not wonder that they become so enthusiastic over its properties that they are able to imagine it the sole reality in existence. For it is certainly the reality in which we are most at home, and it appears to dominate our earthly existence.

Familiar as we are however with matter, we do not really understand it. We do not know its intimate nature. We have only recently learnt that it is composed of electricity; that is to say, that it is built up of unit electric charges. That is a step towards understanding something about it, but we do not know how the electric charges are constituted. We have still very much to learn. Physical Science strongly suggests —though it has not yet proved—that an electric charge is a modification and sensible manifestation of a portion of the ether. If we make the hypothesis that we ourselves are even now more closely and directly in connexion with the ether than with matter, and that our operations on matter are indirect and conducted always by etherial forces, there is nothing in any of the facts known to us which contradicts that hypothesis: and there are some which are rendered easier of understanding when the hypothesis is made. Every kind of physical action is really transmitted across space—that is through the ether—just as really, though not so obviously, as electric and magnetic attraction, gravitation, and light. Atoms and their constituents are never in contact. Ether forces or ether strains have to be appealed to, when we try really to understand the most ordinary activities in daily life. Even a simple push is exerted through an infinitesimal layer of ether. Every variety of potential energy exists in the ether: matter has no energy except kinetic, and recently an etherial explanation of even that kind of energy shows signs of emerging from the theory of relativity.

NEWTON'S VIEW OF THE ETHER

With this preliminary exposition, I want to try to make more generally known what the Science of Physics has taught us about the relationship of matter to ether, leaving the further and more difficult question of the relation of either to life and mind out of the discussion for the present.

The idea of an ether arose in the mind of Sir Isaac Newton because of the difficulty he felt about comprehending the action of one body on another across empty space. He could not conceive it possible that the earth attracted the moon, or the sun the planets, without some intermediate or connecting entity. He considered that physical action at a distance, whereby one body attracted another without any sort of connecting link, was inconceivable. And so he conjectured that all space was filled with a substance of whose properties at that time he knew nothing, but which he conjectured must have to do with the transmission of light as well as of gravitation, and doubtless of electric and magnetic attraction too. Here are his words:—

Query 19. Doth not the refraction of light proceed from the different density of this ætherial medium in different places, the light receding always from the denser parts of the medium?

In Query 21:— . . . if the elastic force of this medium be exceeding great, it may suffice to impel bodies from the denser parts of the medium towards the rarer, with all that power which we call gravity. And that the elastic force [we should now say the elasticity] of this medium is exceeding great, may be gathered from the swiftness of its vibrations.

Query 22. May not planets and comets, and all gross bodies, perform their motions more freely, and with less resistance in this ætherial medium than in any fluid, which fills all Space adequately without leaving any pores, and by consequence is much denser than quicksilver or gold? And may not its resistance be so small as to be inconsiderable?

And again I extract the following disjointed sentences from their context:—

This Æther (for so I will call it). . . for I do not know what this Æther is . . . the vibratory motion of the æthereal medium.

It is true that in Query 28 he gives reasons for rejecting such a medium, partly because its great density would, he thinks, make it obstructive to motion. So that when later he is talking of a possible transmutation of light into matter, he has probably returned to the corpuscular theory of light.

Our knowledge of the properties of the ether may be said to have begun with the undulatory theory of light, early in the last century, when that great subject was being worked at by Huygens, Thomas Young, Fresnel, MacCullagh, and others. But before that, one vital fact was discovered about it through the genius of the Danish astronomer Römer, who—before due time, as it were, and in face of much scepticism,—ascertained that light did not flash across the heavens instantaneously, but was transmitted at a definite and measurable pace. This fact, so easy to specify and yet involving such momentous consequences, I find was known to Newton, who seems to have accepted Römer's result before the end of his life, though *possibly* not till after it had been confirmed beyond reasonable controversy by another result discovered by the English astronomer Bradley.

OLDER VIEWS

This fact about the ether, namely that it transmits waves at a perfectly definite and measurable pace, proved that it had physical properties and belonged to the physical universe. But even now we do not know clearly and distinctly the full nature of the properties which enable it to transmit waves at that pace and no other.

It was found later that the ether in the interstices of transparent matter was in some way affected, interfered with and incommoded, by the presence of matter, being, as it were, loaded with inert material, to an extent which considerably reduced the velocity of wave-transmission; so that the velocity of light inside transparent matter comes down from its proper rate in free space to three-quarters or two-thirds or a half or even two-fifths of its value; according to the optical density of the transparent material. [e.g. (in order) Water, glass, heavy glass, diamond.]

Before that discovery—which was finally clinched by Foucault experimentally, though it was theoretically expected by Thomas Young (1807) and other great geniuses long before,—the only certain and intimate interaction between ether and matter was the fact, known to and clearly expressed by Newton, that ether waves entering opaque material are converted into the molecular agitation which we call heat. And we may consider that the correlative fact was also known to him, viz. that when molecules were agitated sufficiently, that is when a body was made red or white hot, they cause vibrations in the ether which travel out thence in all directions with the speed of light. Hence it is evident that somehow, in a way we now know is connected with what we call electric charge, ether and matter are not independent of each other, but interact. A great mass of knowledge about details of their interaction has now been obtained, though still there is plenty more to learn.

MY PHILOSOPHY.—PART II

The question forced itself on the attention of the French genius Fresnel and the Irish genius Mac-Cullagh, as to *how* matter could react on ether in such a way as to retard or diminish the speed of light inside transparent bodies. It must do this, according to the wave theory of light, and that was sufficient for Fresnel, although the fact had not in his lifetime been verified. But no one knew how this reduction in speed could happen, nor what it was due to. Without going into details, which are somewhat elaborate, it must suffice to say here that Fresnel made the hypothesis that the ether was *denser* inside matter than outside, that is to say that in addition to the free ether which existed everywhere, in space and in matter and everywhere else, there was an additional quantity of what he called "bound ether" associated with and travelling with every material body; this extra portion belonging to the body, as part of it, and apparently inseparable from it. It is not easy to formulate his theory with any precision, and it is doubtful whether ether can be really denser in one place than another. We might prefer to think of it as *loaded* by the atoms of matter, so that it became rather more sluggish than it is in free space. But anyhow, the loading, or whatever it may be,—and to explain it we should have to go into electrical theory and not be perfectly clear even then,—simulates the effect of extra density. That is a mode of expression which must suffice till we know more. And hence the phrase "bound ether," though not unobjectionable, does represent a reality of some kind. There is a certain amount of ether associated with matter, in addition to that universally distributed throughout space. And if the matter is moved, if for instance we were contemplating a stream of water, the bound ether alone would travel with the stream: the free ether would remain absolutely unaffected.

OLDER VIEWS

That, in brief terms, was the gist of Fresnel's theory, and he was able to specify hypothetically how much of the ether was bound and how much free. He said, virtually, that if the whole of the ether inside water (for instance) was divided into sixteen parts, or was expressed by the number 16, then 9 of those parts would be free and 7 of them would be bound. This sounds arbitrary, but of course a clear physical reason can be given for the prediction; and it is given in any relevant treatise on Physics.

After Fresnel's death (that notable genius only lived from 1788 to 1827), another French physicist, Fizeau, made an elaborate experiment to test this theory of Fresnel's. In the year 1851 he caused a stream of water to flow rapidly through tubes, and by ingenious devices he split a beam of light, sending half of it with the stream and the other half against the stream, and then caused the two halves to interfere in a way which would enable him to see whether the half going with the stream was accelerated, and whether the half going against the stream was retarded. The experiment was of course skilfully and metrically arranged, so that he could tell pretty exactly how much was the acceleration and how much the retardation for any given velocity of the stream.

His result absolutely confirmed the numerical result of Fresnel's theory. The fraction came out in accord with it; 7/16ths of the speed of the water is added to or subtracted from the aqueous velocity of light. Others have repeated the experiment since, notably Michelson of Chicago and Zeeman of Amsterdam, with great accuracy; and there is no doubt about it. Forty-four per cent of the ether inside water is bound and travels with the water: the remaining fifty-six per cent is free. A different fraction would apply to glass, and the precise fraction would depend upon the kind of glass. For common glass the fraction would be, not 7/16ths, but 5/9ths. And so on for any substance, in rough relation to its density:—in close relation to the square of the refractive index, just as Fresnel predicted.

If the total ether density in the substance of a body is n^2 times what it is in free space, where its density may be taken as 1, then the free ether, whether inside or out, is not affected by the matter or its motion at all; but the extra, of density n^2-1, is bound closely to the matter, fully belongs to the matter, and travels everywhere with it. This theoretical result may be otherwise worded, but the fact represented by any suitable form of words is and has long been definitely established.

I myself think that the ether is incompressible, and therefore that it is not really denser in one place than another: I am beginning to think that an electric charge by possessing energy in the form of an electric field has diminished the circulatory or magnetic energy in that portion of ether and thereby reduced the vortex elasticity required to transmit waves, but I fully admit that every material body has a specific modification of ether inevitably associated with it; so we may well call it "bound". None of this depends on the body being alive or capable of living. It is true of all matter; solid, liquid, or gaseous, vitalised, or inert.

Now the human body is about the same density as water, as anyone can tell by floating in a bath. Hence we may say that inside the human body, permeated as it is with ether, about half of the ether is free and the other half bound. So far we are on safe ground. But if we go further and ask "Is the bound portion animated?" we are plunging straight into speculation. We are outside the region of physics. We are simply trying a working hypothesis, a procedure which is only justifiable if we are acquainted with facts which will be elucidated by such an idea.

OLDER VIEWS

To me the facts studied in psychical research do require something of the kind for their elucidation: and although we now have to use vague language, it does seem to me possible that some small fraction of this body of bound ether may be that on which we normally and habitually operate, without knowing what we are doing. Also that this fraction of the whole may be, in some unknown way, both individualised and persistent; in which case it would survive the material body to which it is bound, its individual character and persistent identity having been attained in virtue of its being animated, let us say, by Spirit. Animation certainly makes some difference to a body. A live thing can act in a way impossible to a dead or inanimate thing. Then let us suppose that the animation has affected the ether too. If life acts on matter through the ether, it can hardly avoid making some difference to the ether also.

CHAPTER XV

THE ETHER AND RELATIVITY

Although I do not wish to become too technical I will quote a portion of what I said to the Physical Society in 1924, in the hope of interesting some of the younger physicists and getting them to think of a vortex constitution of the ether, until they are able to deduce its properties or to show that they would follow from it on a reasonable hypothesis which I had not the power of working out.

There is a temptation nowadays to assume that the Theory of Relativity has dispensed with the ether. If this only means that the name and the idea need not be used in working out the consequences of certain equations, it is perfectly true. Or if it means that it need not be thought of in writing down or inventing those equations, that also may be stretched as being true—but it requires a good deal of stretching. You cannot write down the Larmor-Lorentz transformations without introducing the velocity c. And, apart from something possessing that velocity, c has no meaning. The usual mode of arriving at those transformations is to think of light as bringing information about events, lagging on the journey and arriving late, and therefore giving us rather confused information about what happens to railway trains and embankments. These popular methods of explanation I mistrust. They inevitably lead one to ask what light has to do with it; why sound or a messenger-boy should not be used

THE ETHER AND RELATIVITY

instead; and absurdities of that sort. I believe that when these "transformations" were first written they were written merely as a mathematical device for dealing with questions of relative motion. But Einstein virtually perceived that there was an absolute velocity in the universe, a velocity c, which was not infinite, and which therefore intruded into the whole of Physics, and could not be dispensed with, except when working approximately. And, however he regarded this velocity c at first, I think that both he and Eddington have come to recognise that if the symbols are to be dealt with not only algebraically, but in a rational and physical manner, there must be something which has this velocity as an essential part of its constitution, and that this "something" must extend throughout the universe, and be responsible for the conveyance of light, electric and magnetic attractions, gravitation, cohesion, and many other forces. In other words, that there must be an ether, and that probably the velocity c represents its extremely fine-grained vortex circulation.

It had already been shown by Kelvin and FitzGerald that, on certain assumptions—some of them rather difficult at the time, and probably none of them quite satisfactory—a vortex sponge would have the property of transmitting transverse waves at a speed of the same order, though not necessarily identical, with the speed of vortex circulation. Hence in all probability, Einstein's fundamental velocity c was of the same order as the velocity of light, and might turn out to be identical with it. Experiment has shown that that is true. And so we speak of c as the velocity of light, not because light is of fundamental importance as the unique and only messenger—or because it happens to be the quickest messenger available—but because light gives us the means of measuring experimentally the order of magnitude of the fundamental velocity c.

MY PHILOSOPHY.—PART II

The fact that that velocity enters into all our expressions ought to give us pause and make us want to know
why and how matter is so related to the ether that this
should occur. The fact, for instance, that when com-
bining two relative velocities for a piece of matter, we
should have to write, not $u+v$, but

$$\frac{u+v}{1+\dfrac{uv}{c^2}}$$

The apparently gratuitous introduction of this alien
velocity c is a tacit recognition of the fact that the bodies
are moving through a fundamental medium to which
this velocity c intrinsically belongs.

That simple equation is an eye-opener to the fundamental nature of things. The equation is derived at
once from the Lorentz transformations, which deal with
space and time separately. But I think we have for some
time been learning that we must regard space and time
as abstractions, and that velocity is a more fundamental
entity, more directly apprehended by us, and that it is
through our sense of speed chiefly that we make our
measures of space and time; and possibly that it is
through speed that we got the idea, at any rate of time.
Hence I prefer to treat this velocity equation as simpler
than its space and time components. And it is well
known that the application of this equation gives us all
manner of important things—the Doppler Effect, for
instance, the remarkable result obtained by Fizeau in
1851, confirming Fresnel's prediction about $1-1/n^2$,
and most other things about relative motion. Well,
then, we ought to realise that the velocity c, and the
ether to which it belongs, are fundamental realities,
which can in no way be ignored directly we cease to do
mere arithmetic and algebra; we cannot ignore them

THE ETHER AND RELATIVITY

when we begin to philosophise, that is as soon as we cease to be mathematicians and become physicists.

There is no need to attend to the ether when we write down such equations as

$$ds^2 = dx^2 + dy^2 + dz^2 - c^2 dt^2$$

but c should be always kept in, and the minus sign, too, since time is then imaginary space—not a real fourth dimension of space.

For shorthand purposes it may be permissible to consider $c = 1$, but I don't like it even for shorthand. It is a confusion of dimensions, and may lead us to overlook some relation which is really important. c cannot really equal 1, for it is a velocity and not a pure number. It may easily be one unit of space divided by one unit of time; but to call it 1 is to ignore the units, and to confuse space and time in an unnatural manner. No velocity or mass or force or anything concrete can possibly be a number and nothing else. I suppose everyone admits that, but considers brevity a question of convenience. I only doubt whether sometimes it is not a question of inconvenience.

In considering the relations between ether, matter, and radiation, there are two modes of looking at things. We may attend to matter first and ether second, or we may consider ether the primary thing and matter the secondary thing. The former is the more practical everyday plan; the latter is, I believe, the more fundamental and instructive and philosophic plan. On that view—the primacy of ether—we endeavour to show that ether in rotary or stationary motion can be made to account for all the forms of energy, and all ordinary activities, excluding those of Life and Mind. Ether and motion must be treated, at any rate hypothetically, as constituting the purely physical universe. What the relation of that to the psychic universe may be is a deeper question, which must be postponed, but on which a few people are beginning to work.

Meanwhile we can hope to find that ether and motion account for all forms of energy, including matter as one of the forms, the form most recently discovered, and perhaps hardly yet universally accepted. If there is controversy about it, that is all to the good, because it will lead to thought and possibly to further experiment. We want to know the kind of motion which constitutes a magnetic field, then the kind of motion which constitutes an electrostatic field, and then—or perhaps before that—the kind of motion which constitutes radiation or light. When we have done those three things, we shall have made some headway.

There are probably some already who think they have a solution of one or other; but I myself feel a difficulty about them, especially about the distinction between the two fields. It can be suggested that a magnetic line of force is a circulation loop opened out or expanded. An electric line of force can be a stretched vortex filament —a core round which there is circulation—and electric attraction may be the end suck of this filament. The ether must have some structure, probably a kinetic structure, or it could not transmit transverse waves; and it is becoming fairly clear that what we call, by analogy, "elasticity" in the ether must be due to vortex circulation. At any rate, that is a reasonable working hypothesis; and it is further probable that that circulation represents magnetic energy. We know also that waves are only generated when electrostatic energy is added to and superposed upon magnetic energy.

As regards the nature of matter, our main difficulty is that we do not know how to generate an electron. Electrons are responsible for all our static fields; but we have to take them as given, just as we have to take magnetic fields as given. What we call a magnetic field is the opening out of previously infinitesimal magnetic loops; and what we call an electrostatic field is merely the summation or aggregation of the fields of a number of electrons. We do not really generate either the one or the other. We separate the charges and study the results; we open out the loops and study those results; and we combine the results into electromagnetism and radiation. We open out the loops by making electrons move. We can do that because they are capable of locomotion, and are amenable to force. We can always make things move, and that is about all we can do.

The generation of an electron would be generation of an electrostatic field. The energy of the field is the intrinsic energy of the electron. If we could generate one, it would presumably have to be at the expense of magnetic energy—that is to say, of ethereal vortex circulation. The production of an electric field would naturally oppose the magnetic circulation, in accordance with a sort of Lenz's Law. But kinetic elasticity depends on the circulation. Hence in the neighbourhood of an electron, kinetic elasticity would be diminished.

Refractive index—i.e., the reduction in light velocity —may be accounted for either by lessened elasticity or increased density, or both. Our first thought is naturally of an increase of ether density inside matter; but we are confronted with other facts which suggest that the ether is incompressible. Moreover, if any medium is really continuous, it is not easy to picture how it can be made anywhere denser than it already is. Compressibility is usually an affair of molecules. If we grant that ether is incompressible, either for this reason or on the strength of the Cavendish experiment—a more exact version of Faraday's ice-pail experiment—we must give up the idea of actual increase of density, and take refuge, as Sir Richard Glazebrook long ago did, in the idea of "loading," a loading of ether by material particles. This idea is attractive, but has difficulties of its own, unless every transparent material or molecular substance is necessarily consumptive of energy. As a revolutionary alternative, we may adopt some modification of MacCullagh's supposition of lessened elasticity.

I know that these terms "density" and "elasticity" were intended to apply to matter, and it is doubtful if there are any mechanical terms which rightly apply to the ether. I quite admit that the dynamics of the ether has still to be made out. But the notion of a hydro-dynamical fluid of intimate and extremely fine-grained vorticity is very attractive, and should not be abandoned without a struggle. It has not yet been given a fair chance, and treated with adequate seriousness. Such treatment is beyond me, but not beyond the coming generation.

On a vortex view we have already seen some reason for lessened elasticity inside matter; and we must admit that if we are allowed diminished elasticity, it could certainly account for the refractive index. So, taking the density uniform, the velocity ratio being $1/n$, the elasticity will be as $1/n^2$. The original elasticity being 1, the modification or change of elasticity, caused by the presence of electrostatic charges, will be $1 - 1/n^2$, and this will belong to the matter and move with it.

When an electric field is generated and superposed upon magnetic circulation, the motion to be expected by Poynting's Theorem need not show itself as a motion of the electron, but may be represented, after Faraday's fashion of an extra current, as diminished magnetic circulation. But when the electron is accelerated, part of the motion is represented by the emission of radiation, which must carry away some of the energy. But that energy will be at the expense of the accelerating force, and will not be original intrinsic energy.

THE ETHER AND RELATIVITY

It is doubtful if a free electron can react on radiation at all. An electron free and isolated could neither radiate not absorb it. When accelerated, it can do both; but to be accelerated it must form part of an atom of matter, and be there subject to the unknown condition of the quantum. Quantum considerations are very interesting, but belong to another chapter. They have nothing to do with the ether alone, they enter into its interaction with matter.

MEANING OF LOCOMOTION

Locomotion is not a property of the ether; it is an affair of molecules. The only motion possible to ether is circulation, but the modification of it that we call matter can move from place to place, and thereby possess what we call kinetic energy—an extra, superposed upon its intrinsic constitutional energy. Let us examine that.

When matter moves, that is, when an electric charge moves, its mass—i.e., the quantity of modified ether associated with it—increases, and this increase of mass accounts for the locomotive energy. That is one way of regarding it. On this view, so far as locomotion is concerned, the speed of circulation remains constant. In fact, there is only one absolute speed in nature, and it is unalterable. Locomotive energy is obtained, not by a change in this fundamental speed, but by addition to the modified portion of ether. It is this modified portion which has perceptible inertia, which appeals to us as matter, and which is capable of locomotion. The locomotive energy is in addition to the intrinsic energy of the modified portion; the intrinsic energy remains constant, as energy of electrostatic field.

Another way of looking at it is to say that the generation of the electrostatic field has reduced the magnetic circulation to something below the velocity c, in fact, to the velocity c/n; but that if the charge is in motion some of the magnetic energy is restored, so that the refractive index is no longer n, but something more or less according to the direction of the motion relatively to the light. And as the speed increases, more and more of the electrostatic field becomes magnetic, until ultimately when the velocity of light is reached, the circulation energy is all restored to the speed c, the refractive index becomes 1, and nothing more can be done.

It is not possible to add to the velocity c, but it is possible to add to the velocity c/n. It can be added to until c/n has become c, when either $v=c$ or $n=1$. In the latter case matter will have disappeared, the ether will have become normal and unmodified, and locomotion has no more meaning.

This means a reduction of mass instead of an increase, and is probably wrong. But we might take it the other way, and say that the magnetic circulation becomes more and more converted into electrostatic charge, and therefore mass; until ultimately it is all used up, and the whole of the circulation has become locomotive with the velocity c.

Energy of radiation has locomotion, but only at a definite speed, and it has no intrinsic energy at all. Its energy can be all used up either in accelerating electrons or generating electrons. Radiation is merely a mode of transmitting energy from place to place. It delivers up at one end what it has received at another. But in what form the energy is disposed of, and what remains when its locomotion is taken out of it, remain to be investigated in conjunction with photo-electric and other effects.

The only way we can deal with or gain access to magnetic circulation is by opening up its loops, making it circulate round a finite area. A magnetic field is always a flow in closed curves. But electrostatic energy is different. An electric field is not a closed curve. It always starts from one electron or group of electrons and ends on another; though each line of force may be surrounded by a whirl. There can be no electric field without matter, any more than there can be matter without intrinsic electric fields. Neither can there be radiation without matter, because to produce radiation both electric and magnetic fields are essential. Nevertheless, when radiation is once generated by means of modified ether, it can travel on through unmodified ether, wherein, previous to its arrival, there was nothing but magnetic circulation. In this way the influence of an electric field is propagated from place to place, and an electric disturbance can be caused or reproduced at a distant station when the radiation arrives, just as it was momentarily caused in a temporary manner all along the ray while the radiation was being propagated.

Because of our special sense organs, we naturally attach great importance to the peculiar modification of ether which appeals to us as matter. It need not have any more energy than an equal bulk of unmodified ether, at least not to a first approximation. It may be that there is a second order effect to be discovered, which may be of great importance; but omitting that, matter represents etherial energy in another form, the form we know as electrostatic.

The ether-circulating velocity is always c, except when some of it takes the electrostatic form—whatever that may be—and then it becomes c/n, the electrostatic portion accounting for the difference.

Our main difficulty at present is understanding what is the real constitution of an electric field, which is evidently involved with the unknown constitution of an electron. When one is understood, the other will probably be understood too.

MY PHILOSOPHY.—PART II

Given a great amount of magnetic energy in the ether, given that some of it has already taken the electric form—that portion constitutes matter, and what we call the inertia of matter is the electric energy which constitutes it. What we call the velocity of matter is a quasi and temporary addition to this inertia, that is to say, a slight addition to the amount modified, extra and beyond its original intrinsic value.

CONCLUDING REMARK

From this point of view, as to the nature of matter and its inertia and its locomotion, there is an addition to be made to what is said above about the composition of velocities, that is of the relative velocities and kinetic energies which come under our familiar observation. Every relative velocity is a sense impression derived from some modified portion of ether acquiring extra or at least modified energy. The fundamental velocity c continues unchanged; nothing is added to it, but some of what had been purely circulatory energy becomes locomotive. This is an addition to the constitutive energy, m_0c^2, of the original modified or material portion of ether, for it now becomes βm_0c^2, and the locomotive energy is $(\beta-1)\, m_0c^2$. If this is expressed in terms of v^2, it may be well to express v in terms of c, and in fact to consider v as an abbreviation for v/c. In that case all observed velocity is really a ratio, and we have an excuse for thinking of c as 1. The composition law, for two velocities u and v, is now

$$\frac{w}{c} = \frac{u/c + v/c}{1 + \dfrac{u}{c} \cdot \dfrac{v}{c}}$$

and the c^2 comes into the denominator naturally.

Ideas such as this must not be treated as obvious and ignored. Though in one sense simple, they must really

166

involve a deep meaning. And it is in the hope that the present generation of younger physicists will decipher that meaning that I have set forth my own thoughts on the subject. This chapter may be likened to a kind of radiation which however feeble in itself may, by virtue of its frequency or some other peculiarity, succeed in liberating an unexpected quantum of energy from the friendly and responsive material upon whom it falls.

CHAPTER XVI

MAGNETISM AND THE ETHER[1]

WITH SUGGESTIONS FOR EXPERIMENT

The first thing that has to be said about magnetism is that you cannot generate it. You can generate heat, and light, and sound, but not electricity, or magnetism, or matter. No one expects to be able to generate matter; not that it is hopelessly impossible to imagine, but that it has never been done, and the doing of it would be a great discovery. People did think at one time that they could generate electricity, but we know now that they can only set it in motion or re-arrange it, just as if electricity were a form of matter. Many still think that they can generate magnetism, and certainly they can make any number of magnets, but always by aid of an initial magnet. So also they can generate as many oaks or currant-bushes or chickens as they like, but there is no life without antecedent life. Any twig cut off a currant or gooseberry bush will develop into another bush, but it cannot be generated *de novo*. The specific germ of life has to pre-exist, and then an acorn can produce in time a whole forest of oaks; showing that life is a thing *sui generis*, and not a form of energy, for there is no conservation about it. It is nothing like energy and cannot be converted into any other form. It must have pre-existed and be brought into inter-action with matter.

[1] Part of a Silvanus Thompson Memorial Lecture to the Röntgen Society in March, 1922.

MAGNETISM AND THE ETHER

So also with magnetism. It is always passed on. And the energy belonging to it has to be supplied by the mechanical or chemical arrangement which has apparently produced it. In the case of life the energy is supplied by food; in the case of magnetism, by the muscular or other action which withdrew the magnet from the generating field.

But, it may be said, "What about an electromagnet? Surely that is generated by an electric current." I admit that an electric current can be generated, by the process of setting electrons in motion; but the evidence goes to show that the magnetism which accompanies an electric current is something pre-existing in the ether, and is not called into being out of nothing. Heat, light and sound are forms of energy. Magnetism is no more a form of energy than is electricity or life. Magnetization may be called a form of energy; so may electrification.

How then are we to visualize the act of magnetization, say, when a piece of iron is magnetized? I say that the act consists in opening out molecular magnetic loops which are already in existence. They are like rings swelling sideways. You cannot cut them or get at their ends, as you can cut a loop of string. They are more like indiarubber bands which can be stretched, stretched without limit; so that, instead of being shut up into an ultra-microscopic space, they swell out so as to enclose a considerable region. The lines of force of an active magnet envelop a large area, and when the magnetism is destroyed or suppressed, the loops do not cease to be, they merely shrink, so as to enclose a smaller and smaller area as the magnetism dies away; and they remain, ready to be swollen out again when the magnetism is re-excited.

Matter is not necessary for a magnetic field. You can magnetize a vacuum. Air has nothing to do with it.

MY PHILOSOPHY.—PART II

But iron and other magnetic substances have a large number of such loops in their own constitution, so that when these are opened out they add themselves to the loops which exist independently of matter, and thus produce a stronger field for a given magneto-motive force. But such extra loops are not essential, for nothing has yet shown that an equally strong magnetic field cannot be produced in empty space,—that is in the ether,—if a sufficiently strong current is available. And indeed the field surrounding the equator of a rapidly moving electron is of prodigious intensity, estimated as 10^{15} c.g.s. units for an electron moving at one-thirtieth of the speed of light. This intensity of field is still greater as the velocity of light is approached; and then there does seem to come a limit, when the resources of the ether are exhausted, so that it becomes impossible to make an electron move at more than the speed of light.

Besides these tremendous circular fields there are the loops which can be opened out in the process of magnetization; but even when opened out the loops themselves remain closed curves, passing right through the core of the magnet and never terminating on its surface in the way that electric lines of force terminate.

An electric current is also continuous and circuital; but when the current stops, the portion inside a conductor ceases to be, and the residual lines of electrostatic-strain stretch only from one conductor to another, through the intervening insulating region; their terminations being what we know as opposite electric charges. And these opposite charges can be moved apart from each other and neutralized, in a way that does not happen with North and South magnetism. North and South magnetic poles cannot be separated from each other. If a magnet is cut, you get two magnets. The lines are found to have threaded through the material, and you only expose the poles which were already there. The poles, in fact, are what we observe when a magnetic line of force passes from one medium into another. We observe that also with electric lines, but in the interior of a conductor the electric lines have ceased to be. That is not so with magnetic lines. They are necessarily and always closed complete curves.

A current loop and a magnetic loop are always correlated, they are linked together and one seems to be just like the other. We know that there is circulation round the one. There may be circulation round the other. But there is this difference: the current circuit can be cut, and when the current stops the circulation stops dead,—stops in its tracks,—leaving, if any residue at all, an electrostatic field between the cut ends; but the magnetic loop behaves differently. When the generating current strengthens, it expands; when the generating current weakens, it contracts; and when the generating current stops, the magnetic loop shrinks into nothingness; not into nothing, but so as to enclose only an infinitesimal and therefore imperceptible area. The Ampèrian circuits remain, to be opened out afresh when occasion arises; and these circuits, though they exist in matter, exist also in empty space; so that primarily magnetism is an affair of the Ether.

POSSIBLE CIRCULATION

Now when we find a circuit necessarily and always closed, there is an obvious possibility that something is circulating round it. We know something of what is circulating in an electric current; we do not know what is circulating in a magnetic loop. The theory of Sir Joseph Larmor and others, which has some kinship with the vortex theory, postulates that in all probability the ether of space is the thing that is circulating; for the ether, although absolutely stationary as regards locomotion, and stationary also with respect to every mechanical activity, need not be rotationally stationary. It may be capable of spin, in closed circuits, though it never seems to move from place to place; it may go round and round like a vortex. And many facts, in Electricity and Magnetism and Light, suggest that this hypothetical circulation has some foundation in reality.

The electric properties of the ether suggest something akin to elasticity. The magnetic properties of the ether suggest something akin to inertia. And, by the interaction of these two properties, wave propagation is rendered possible; whereas except for those two properties no wave propagation could occur. The two properties of the ether, though at present unknown, are designated by the letters K and μ. One is the electrostatic property, and the other is the magnetic property. They are quite fundamental and universal, and it is a great mistake to ignore them or to consider that either of them may legitimately be called unity; except for practical purposes, to cover our ignorance until we know something about them. Maxwell showed how to determine their product. He proved, for the first time, that the reciprocal of their geometric mean was the velocity of light or that $\mu K c^2 = 1$. That is all we really know for certain about them. It is highly desirable to know more; and, by keeping them constantly in mind, it may be hoped that the present or some future generation will devise an experiment whereby their ratio or some other function of them can be determined; in which case we shall know both, and begin to have a real science of the ether of space.

Or it may be that an experiment can be devised capable of measuring either of them separately, which would serve equally well, for then the other would of course be known. At present we know neither their nature nor their value, but we can make hypotheses; and on the strength of a working hypothesis we can try experiments to see whether any result can be got. The working hypothesis on which I proceed is to suppose that what the mathematicians have thought likely is true, and that there is a slow etherical circulation or flow along the lines in a magnetic field,—a flow which cannot be excited or stopped, but which can be opened out so as to be made perceptible,—the flow being no longer in infinitesimal orbits but in orbits of considerable size, so that a beam of light may be sent along or against the flow and a measurement be made to see whether the light is accelerated or retarded.

I have spoken of the circulation as "slow," for so I think it is, in any practicable magnetic field: that is because of the enormous density of the ether. The rate of flow would have to be estimated from a knowledge of this density, and of the energy of the magnetic field; for the energy of the flow, and the energy of the magnetic field, would be one and the same.

Now the energy of any given magnetic field can be easily calculated; hence, if we know the density, we know the rate of flow. Or, if we can experimentally measure the rate of flow, we can thereby determine the density of the ether, or of whatever it is that is flowing.

Nowadays it is becoming customary or fashionable in a few quarters to doubt the existence of the ether of space, and to suppose that Einstein has exploded it. That is contrary to the truth; and, so far as I know, no great authority on Relativity really supposes that; certainly Einstein does not himself, nor does Eddington. On the contrary, Einstein's discovery, linking gravitation with light for the first time, strengthens the position of the ether. For light has always been known to be an etherial phenomenon, involving everything that is requisite for wave propagation,—in light, of course, I include X-rays,—and Einstein has now linked gravitation with the ether too, since he has connected it with the phenomenon of light.

Hence Relativity, though it uses different language and need not mention either ether or matter, really strengthens the position of those who hold that a medium is necessary for the propagation of activity from place to place. What is called "Action" essentially involves continuity and a medium.

Admittedly, however, a flow of ether along lines of magnetic force is a working hypothesis, not yet justified however by any clear demonstration or crucial experiment.

IDEA OF A SPECIAL EXPERIMENT ON ETHER CIRCULATION, BY MEANS OF LIGHT TRAVELLING ALONG A MAGNETIC FIELD

Many years ago, in fact during the nineties of last century, I myself made an elaborate experiment seeking to determine the speed of the hypothetical magnetic flow, and thus reach a lower limit to the density of the ether. I could have detected a flow of a foot a second, in the strongest magnetic field I was then able to produce without iron and over a considerable length. But, in subsequent years, and finally in 1907, I made estimates of the probable density of the ether, which, if they are considered justified, would prove that the flow I was looking for was vastly slower than anything I could then hope to detect. I would have had to find something of the order of a few inches an hour, less than a snail's crawl—a glacier-like speed; a thing very difficult to compare experimentally with the velocity of light. An experiment on these lines therefore is almost hopeless,—almost but not quite. I think that with refined experimental skill and unlimited material resources it might be done, and I wish to make an appeal to Dr. Kapitza, at the University of Cambridge, to undertake such an experiment.

Meanwhile I propose to give to this Society some account of my own attempt, for it has never been properly published; indeed recently Larmor has urged me to make it better known, since though negative he regards it as important so far as it went. Negative experiments always seem disappointing, but occasionally they turn out very important. I need only instance Michelson and Morley's great experiment on ether-flow near the earth. The result was negative, but it was the basis of the FitzGerald-Lorentz contraction, and has been made the foundation for the mathematical Theory of Relativity. I emphasize this, because it is really a remarkable consequence to follow from a negative result; meaning by a negative result not anything vague or indefinite, but a definite answer, nought or zero, to a question as to the value of something.

If, however, a positive ether flow could be demonstrated by sending light along the lines of a magnetic field, the consequences would be enormous. Our knowledge of the ether would increase by leaps and bounds. The energy which has been hidden away in its internal ultramicroscopic or rather infra-atomic finely-grained motions would be displayed, and this energy would be found to be of an order altogether overpowering the energy locked up in the atom; for it is as much greater than any atomic energy as that atomic energy is greater than those sources on which we now depend, such as water power and the combustion of coal.

All our energy comes, even now, through the ether from the sun; but it has had a material origin or source, in the solar orb, and it may have been excited by the atomic—as distinct from molecular—processes there going on. In that sense it may be said that we are already tapping some of the energy of atoms; but no idea has yet dawned of tapping the intrinsic energy of the ether, which, if my hypothesis is right, is equivalent to the output of a million-kilowatt power-station working continuously for several million years, locked up in every cubic millimetre of space.

The universe which we are only just beginning to explore is of a portentous character. The rate of vibration which constitutes ordinary light, and which our eyes are adapted to make use of, is of an incredible magnitude. No one can conceive of a vibration of five hundred million million per second, and yet that such vibration is a reality is not only a certainty but a commonplace of science, and has been known for a century. So the fact that large numbers and high values are involved ought not to be a deterrent. We have to ascertain what is, and not assume that anything is *a priori* impossible. Nor need we be astonished at the almost infinite possibilities which are now looming before us in many directions, some of them not yet incorporated but looked at askance by orthodox science; in some of which, by the way, I may mention incidentally, that prolific experimenter, Sir William Crookes, was also a pioneer.

It is rather odd, perhaps, that the two first honorary members of the Röntgen Society should be both tarred with the same brush.

To give some idea of the importance of some such experiment, and some notion of the reason for expecting any result, or at any rate for the theoretical implication of a result, either positive or negative, the following extracts from letters received from Sir Joseph Larmor may be quoted:—

MAGNETISM AND THE ETHER

> "28 Oct., 1893.
> "Your kindness in sending me a copy of your (*Phil. Trans.*) Aberration paper encourages me to ask you a question which now interests me very keenly, namely, whether transmission along lines of magnetic force in air appreciably alters the velocity of light. I cannot discover that anyone has tried, but according to an etherial theory which I have nearly finished writing out, and which covers, I fondly imagine, a very extensive ground, there ought to be such an effect. Perhaps you know, or could easily try with your machinery."

I suppose I replied that I had already tried in different media, and had found nothing, but that I would repeat the trial more carefully if he would be satisfied with an experiment in air; for his next letter says:—

> "1 Nov, 1893.
> "I can correlate most things in one scheme if I am allowed that magnetic force is velocity of the ether. . . . A magnetic field should carry the light along with it,—with what kind of velocity I have not yet formed an opinion; but it has nothing to do with rotary effect, and should be of the same order in vacuum as in a mass."

I now turn to a description of the series of experiments made by me between 1893 and 1897, over and above those published in the *Philosophical Transactions* for those years; though they were partially described in the *Philosophical Magazine* for April, 1907, as a sequel to calculations about the probable density of the ether.

The arrangement involved an electrical part and an optical part. The electrical part consisted of four brass bobbins or frames, each 44 centimetres long, with an elliptical core 10 centimetres wide and 1 centimetre high. This frame was provided with cheeks, so that a great deal of wire could be wound upon it; but in order to keep the wire cool, the central portion and the cheeks were made double, so that there was an interspace which could be filled with water, the water being kept flowing rapidly by connection with the mains. Thus a film of running water was everywhere interposed between the wire on the bobbin and its central core, where the light was intended to pass. Each bobbin was wound with 7,000 turns of silk-covered No. 18 wire, in twenty-two layers, and when the four bobbins were put in parallel, and 230 volts applied to them, the total current was 28 ampères, thus expending about nine horse-power in heating the wire, so that it was inexpedient to leave the current on for more than a few seconds, which, however, was amply sufficient to make the optical observation. Each bobbin was provided with glazed ends, of optically-worked plate glass, that could be clamped on in a watertight fashion, so as to introduce any desired liquid into the core. When all the bobbins were placed in a row, and filled with bisulphide of carbon, the rotation of the plane of polarization of sodium light, with a current from 110 volts, was 83°; and from this datum, as well as from the ampère-turns, the total drop of magnetic potential could be estimated.

The bobbins worked well, and the heat-consuming arrangement was satisfactory for the case of air and water, but bisulphide of carbon is so extremely sensitive to changes of temperature when tested optically, that either weak currents had to be used for a specially short time, or this material had to be discarded. There was no particular object in employing a dense material, and it was a great relief to fall back upon air, so as to dispense with the glazed ends of the bobbins, and thus avoid a multitude of reflections. Even air, however, was to some degree sensitive to changes of temperature, and if I had pursued the experiment further, I should have been tempted to restore the glazed cover-plates, and use vacuum. But it would have complicated matters a good deal, and did not seem necessary. I wanted to try a

MAGNETISM AND THE ETHER

magnetic liquid, but unfortunately no magnetic material is transparent. The discovery of a transparent magnetic substance would be of great assistance to magneto-optics; but it may be that such a substance is impossible, though I know no precise theoretical reason for such impossibility. A transparent conductor of electricity, other than electrolytic, was shown by Maxwell to be impossible; but a transparent substance of high magnetic permeability might perhaps conceivably exist, though I am very doubtful if it can. Anyway, I was limited to the poor permeability of salts of iron, which I sometimes used in spite of the rusting and other trouble they gave; but their results were of no particular interest, and my best experiments were made with the cores merely containing air.

The optical arrangements were of an elaborate and very careful kind. They were the same as had been used in what I may call my great ether experiment, made during those same four years, and described and depicted in the *Philosophical Transactions* for 1893 and again in 1897. The same stout oak frame about a metre square, and the same optically-worked mirrors made by Hilger, and the same telescope and micrometer which had been presented to the laboratory by Dr. Isaac Roberts, F.R.S., were employed.

Light from an electric lantern was sent first on to a piece of semi-transparent silvered glass, so as to divide it into two equal portions by reflection and transmission. Plane mirrors, supported in geometrical fashion and accurately adjustable by screws without strain, were supported on the inside faces of the square, so as to catch the half-beams of light and send them round and round the square in opposite directions, sometimes once, more often three or four times, and occasionally five or six times. The arrangement was such that they ultimately met again on the semi-transparent glass and

179

entered the object-glass of the telescope, thus producing interference bands, which could be seen with great precision in the eyepiece. The eyepiece was provided with two micrometer wires, both movable by separate micrometer screws: and the usual plan was to set one wire—the straight one—on the middle of the system of interference bands, and adjust the other one in some well-marked feature, say the yellow, of one of the other bands. In this way it could be seen if the system shifted as a whole and also if the bands widened or narrowed when the magnetic field was applied. The bands could be obtained quite reasonably broad, and the 1/100th part of a band shift could be estimated. Some skill and experience were necessary to get these optical arrangements perfect, especially when the light had to pass through a considerable column of liquid bounded by glass plates, but the perseverance of my assistant, Mr. Benjamin Davies, overcame all these difficulties; and indeed he had already had plenty of experience of this kind when the whirling steel discs were inserted in the optical square, in the main ether experiment, instead of the magnetized bobbins. That experiment with the whirling discs was much more difficult than the magnetic experiment now referred to, for the blast of air thrown off the discs when they were spinning like a teetotum under the impulse of an electric motor directly on the shaft,—nine horse-power being expended in driving them,—entailed all manner of precautions, and some elaborate devices. The magnetic experiment with the four bobbins entailed no such difficulties. The only real trouble was the inequalities of temperature, inevitable even with the fairly efficient water-screen when the current was left on too long.

Suffice it to say, therefore, that the four bobbins were arranged as the sides of an inner square, at 45° with the main optical square; that light was sent through the bobbins three or four times both ways; and that a key was provided for the observer at the eye-end of the telescope, so that when the interference fringes were in good condition and the micrometer properly set, he could switch the current on, stop it, and reverse it, at pleasure, noting the effect, if any, on the bands.

It will readily be understood that when a liquid was in the core, so that the light had to go through it and through the covering glass plates, the slightest shift of the bobbins was sufficient to cause some disturbance, especially since in this arrangement of a square, instead of an equilateral triangle, the going and returning light did not traverse identical, but only parallel, paths on their way round and round the square. That is a defect of a frame with an even number of sides, and it would be avoided by using an odd number, for then the opposite paths of the light would be identical, making the arrangement much less sensitive to perturbations en route. But then an odd number of sides would not enable the light to be sent many times round: it could only be got once round, thus reducing the effective length of path. I do not say that all this is obvious: it will be found so on drawing the diagram fully. The effective length of path in my case, when the light went four times round, was 7 metres for each half-beam, or 14 metres in all, and in this length of beam an acceleration or retardation of as little as the $1/100$th part of a wave length should have been observed when conditions were good.

Spurious shifts, of course, had to be avoided; the magnetic attraction of the four bobbins was quite sufficient to cause spurious shifts when there was a liquid in their cores; and there were other sources of disturbance, which can readily be imagined by those who have tried delicate experiments. Nevertheless the result is very definite, the difficulties were by no means insuperable, and when only air was in the core, were not really very great. And the final result is that no shift was seen, in spite of the fact that the total effective drop of magnetic potential was two million c.g.s.

The conclusion from this is either that the theory about longitudinal magnetic drift is erroneous or else that the ether of space is at least eight times as dense as platinum. For if the energy of the magnetic field is due to ether-flow, it is manifest that the denser the ether is, the slower it need flow. The rate of flow which I could have detected, along a field of intensity of 1,400 c.g.s., was one foot per second. I can therefore safely say that if there is any etherial circulation along the lines of a magnetic field of intensity 1,400 c.g.s., it is less than one foot a second.

But after all one foot a second is rather rapid, and on the estimate of ether density which I made in the same *Philosophical Magazine* paper of April, 1907, where on page 492 I calculate the massiveness of the hypothetical magnetic circulation, the ether density comes out 10^{12} and with such a density as that—which is some thousand million times the density of platinum—no flow of anything like a foot a second can be contemplated, except indeed close to the equator of a moving electron, or in some other ultra-minute closed contour, such as we are not likely to be able to open up into a perceptible or experimental magnetic field. All we get in our magnetic field is a slight residue merely of what must be going on with tremendous energy in the innermost nature of the ether.

It becomes therefore a question whether it is possible to make a magnetic field of such strength, and to have all equal arrangements of such perfection, that any ether flow producible on a large experimental scale could be detected. I have written on this subject in the *Philosophical Magazine* for May, 1919, and I there reckoned that if the four bobbins, above described, were wound with a thousand turns on every centimetre of effective length, and if, say by liquid air, the wire were kept so cold that it could transmit 1,000 ampères for say, a couple of seconds, and if the bobbins were 10 metres long instead of only about half a metre, it would be possible to detect an ether-flow of about 7 inches a minute, which is about the order of flow that might theoretically be expected from an ether density of 10^{12}.

The importance of the experiment is that if we obtained a positive result, that which we call the density of the ether, or in other words the magnetic constant $4\pi\mu$, would be determined, and then the elastic constant $4\pi/K$ would be known too, for Maxwell showed long ago—actually as long ago as 1865—that the ratio of these two (thus specified) great etherial constants was the square of the velocity of light. Hence, if one is known the other is known also, and a third step will have been made in our knowledge of the properties of the ether. The first step was the determination of the finite velocity of light, a vitally important thing, bringing the ether definitely into Physics, and showing that it had finite and measurable properties. The second step was taken when Maxwell showed that this velocity was dependent on the ether's electric and magnetic properties, and that it could be measured by purely electromagnetic experiments, without making use of light in any way except to see the instruments. This, as is well known, gave us the measure of the product μK. And now the third step awaits the enterprise of some of those of the younger generation who undoubtedly one day will devise a method and make the necessary, probably difficult, experiment, to determine some other function of these two constants, or to determine one or other of them separately.

MY PHILOSOPHY.—PART II

At present we have no certain knowledge even of their nature, still less of their magnitude, and in consequence of this ignorance there are two systems of measurement in vogue, one the electric, which assumes that K is 1; the other the magnetic, which assumes that μ is 1. Both assumptions are known to be absurd, and are only justified by practical necessity. From the practical point of view the justification is complete; for except on some such convention, the electric and magnetic units of measurement would have been impossible. But physicists should vividly remember that they are based on nothing but a convention, and should keep the two unknown constants of the ether present in their minds, in the hope some day of simplifying the whole matter, and removing the unnecessary, though temporarily inevitable, complication.

When I say that the two constants are absolutely unknown, I do not mean that they cannot be speculated upon. As a matter of speculation, I suggest that μ is of the nature of density, or analogous to something which in the case of matter we call density, and that it will be found to be of the order 10^{11} c.g.s. units. Also that K is of the nature of the reciprocal of an elasticity, and will be found to be of the order 10^{32} c.g.s. units. But it is quite possible that posterity will jeer at these estimates, if it ever knows anything about them. What I do not expect posterity to jeer at is the attempt to think of some plan whereby their nature and value can be discovered; and I hope that they will regard with lenience the effort, however tentative and mistaken it may be, to make an estimate on the basis of present knowledge of what really requires some fundamental and important discovery.

As this is a Silvanus Thompson Memorial Lecture, in which the history of Science has its due place, I propose to conclude with the last page of my *Philosophical*

MAGNETISM AND THE ETHER

Magazine paper of May, 1919, as a kind of summary of what the main part of this lecture has been about. Referring to the experiment above roughly described, I first say that "if a distinct answer can be gained, the experiment is well worth while."
And then I go on:—

"Perhaps it is not clear why I attach so much importance to a measurement of the etherial constants and a determination of their dynamical nature. If a positive result could be secured it would be the first positive result which the ether, apart from matter, had yielded since the fundamental fact of wave propagation and its definite velocity. This determined the product of the two constants, and was the first step in our knowledge of them. If, however, by some new phenomenon the two constants could be separately known, a second and even more important step would have been taken towards understanding the ether's structure and real nature. Until these constants are known, its relation to and interaction with ordinary matter must be largely guess-work. Radiation, once excited, obeys known laws, but of the emission and the absorption of radiation very little is really understood; and even the refraction or slowing-down of speed when passing through dense matter appears to be a subject of some difficulty, at least when anything more has to be apprehended than the bare fact and its elementary exposition.

"If the density were known, of course, the elasticity would be known too, unless the dynamics of ether is not merely a variation on Newtonian dynamics but something utterly different. The only way to ascertain the truth on this subject is to try how far the ether can be treated as a substance amenable to ordinary laws. The principle of Least Action holds for light, and it seems possible that a developed turbulent or vortex sponge theory may account for the ether's elastic rigidity (cf. Appendix E, and p. 124, of Larmor's *Ether and Matter*. Also *Philosophical Magazine* for April, 1907, p. 503). It is essential, however, that we know the value of this rigidity. If it is kinetically explicable in the way originally suggested by Lord Kelvin (though afterwards abandoned by him) then the amount of energy locked up in the ether is something prodigious. Some day such a fact as this, when ascertained, may be found to have a bearing on really practical problems."

185

CHAPTER XVII

SUMMARY OF OUR PRESENT KNOWLEDGE ABOUT THE ETHER

To conclude this portion I will append an Article written for the 13th Edition of *The Encyclopædia Britannica* in which I sum up what is known to orthodox physicists on the subject at the present time.

ETHER (in Physics)

Whether space is a mere geometrical abstraction, or whether it has definite physical properties which can be investigated, is a question which in one form or another has often been debated. As to the parts occupied by matter, that is by a substance which appeals to the senses, there has never been any serious doubt; almost the whole of science may be said to be an investigation of the properties of matter. But from time to time attention has been directed to the intervening portions of space from which sensible matter is absent; and this also has physical properties, of which the complete investigation has hardly begun.

These physical properties do not appeal directly to the senses, and are therefore comparatively obscure; but there is now no doubt of their existence, even among those who still prefer to use the term "space." But a space endowed with physical properties is more than a geometrical abstraction, and is most conveniently thought of as a substantial reality, to which therefore some other name is appropriate. The term used is unimportant, but long ago the term ether was invented; it was adopted by Isaac Newton, and is good enough for us. The term ether therefore connotes a genuine entity filling all

186

SUMMARY OF PRESENT KNOWLEDGE

space, without any break or cavity anywhere, the one omni-present physical reality, of which there is a growing tendency to perceive that everything in the material universe consists; matter itself being in all probability one of its modifications.

Many attempts have been made to state the properties of such a substance in terms of material analogies, and all these attempts have shown signs of weakness and may be said to have failed. The properties of the ether are too fundamental to be stated in terms of something else.

There have been tendencies at different times to invent ethers or effluvia with special qualities to account for specific phenomena. These attempts were long ago discarded, and are now regarded as absurd. But that space has physical properties is a definite fact of experience, provided experience is extended to include inferences and deductions and is not limited to direct sensual perception. What we perceive directly are length, breadth and height, modified here and there by a resistance or obstruction which we call matter, and combined with the element of time or duration as exhibited and measured by the motion of matter, with speeds that can be directly apprehended.

But in addition to all that mass of common experience, the free unobstructed space is modified by the neighbourhood of matter; so that there exists everywhere a gravitational potential varying inversely with the distance from its appropriate portion of matter; the result of which is that matter tends to move from places of higher to places of lower potential, as if some force were driving the masses of matter together. Civil engineer-ing—the erection of structures and the movement of great masses of material—is constantly concerned with this fact; and on this basis the whole of the older astronomy has been worked out in the most intricate detail.

TESTIMONY OF OPTICS

The atoms of matter are not quiescent, even when a mass appears stationary, but are in a state of rapid quivering motion; and these motions are not independent of each other, but are interrelated and connected by additional and special disturb-ances which they communicate to the space or medium in which they occur. And about these supplementary disturbances our sense organ, the eye, has given us a mass of indirect information. These disturbances, though generated by matter, are not conveyed or transmitted by matter. They travel at a rate depending on innate properties of space; or rather, as we feel bound to say, on the physical properties of the substantial reality which fills space; thereby telling us something definite about those properties, though in a form difficult of apprehension, and one which is not fully expressible in terms of any of the familiar properties of matter.

Thus the different masses of matter, even though separated by great distances, are not isolated or independent of each other. They are connected gravitationally, and they are connected optically. The energies of the earth, of which we constantly make use are derived from the sun, and have travelled across the intervening 92,000,000 miles of empty space at a perfectly known and definite rate, with which rate matter has nothing to do. There may be uncertainty as to what exactly it is that is travelling; but the fact that it is travelling energy is certain. All that matter does is to generate this radiant energy at one end and absorb it at the other.

Concerning the processes of generation and absorption a good deal is now known. Moreover not only is the speed of travel of the transmitted disturbance known, but also the fact that it is a periodic disturbance, expressible mathematically in exact analogy with a wave equation. Wherefore the disturbance may be spoken of without further hypothesis as ether "waves," the generic name for which is "radiation," a small range of this radiation being visible light.

Radiation is generated by some cataclysm or collision or other violent and sudden disturbance in the atoms of matter. When radiation encounters matter (unless it be merely reflected or passed on) it can throw the multitude of atoms into the confused motion we call "heat," and produce other remarkable and chemical effects. Thus an ether is necessary for the purpose of transmitting what is called gravitational force between one piece of matter and another, and for the still more important and universal purpose of transmitting waves of radiation between one piece of matter and another, however distant they be.

SUMMARY OF PRESENT KNOWLEDGE

In addition to those two functions, other properties have been discovered, notably the properties called electric and magnetic. Atoms of matter are electrically constituted, and accordingly tend to attract each other with a force which is the source of chemical affinity; with the result that molecules and other aggregates are formed, of which the structure is studied in the science of chemistry. Moreover the molecules themselves attract each other by a residual affinity, giving the familiar shape of crystals and other solids, the particles of which are held together in regular packing across ultra-microscopic intervals by what is called cohesion, for which likewise the ether must be held responsible. For, as Newton forcefully said in other words, it is absurd to imagine one piece of matter acting mechanically on another at a distance, whether that distance be large or small, without some intervening mechanism or connecting link. The continuous medium which fills space, therefore, is not only the vehicle of gravitation and light, but is also the instrument for cohesion and chemical affinity and for electric and magnetic attractions and repulsions. It must also be the vehicle for every kind of mechanical force, and for the elastic connection between particles of matter, which are never in real contact with each other.

The intimate structure of the ether may ultimately be expressible and partially understood in terms of the phenomena of electricity and magnetism: for electric and magnetic influences are transmitted perfectly through vacuum, that is, across space empty of matter. They represent primarily properties of the ether, though they are only made manifest to our senses by means of matter. It was in terms of electricity and magnetism that Clerk Maxwell was able to explain the phenomenon of light. A close study of electro-magnetism, that is, of the interaction between electric and magnetic disturbances, showed that they must combine into a wave equation, the waves being transmitted at a rate calculable from purely electric and magnetic considerations. This velocity turned out to be the velocity of light; and so in 1865 the true theory of light was born.

MY PHILOSOPHY.—PART II

Not that it is anything like complete. We know too little of the electric and magnetic properties of the ether to be able to picture exactly what is happening. What we do know is that light is an electro-magnetic phenomenon, and that it is entirely dependent on the properties of the ether. The ether involves or possesses properties expressible by two fundamental constants; one of which regulates the force of attraction between two electrified bodies, and the other the force of attraction between two magnets. Neither constant by itself is as yet known. But the value of the constants multiplied together is known; it was discovered by Clerk Maxwell, and is the reciprocal of the square of the velocity of light. In other words, the combination of the electric and magnetic properties of the ether enables it to transmit waves at a rate equal to the inverse geometric mean of its two constants.

So far we have been dealing with things which have been known for some time. But the subject is so fundamental and important that a recapitulation in other terms seemed advisable. It now remains to deal with the later progress which has been made in investigating the properties of this extraordinary non-material but physical substance. Perhaps "substance" is hardly the right term, for, though exceedingly substantial in one sense, it makes no appeal to the senses and is therefore unlike any substance we know.

In the 9th Edition of the *Encyclopædia Britannica* an attempt was made to estimate the elasticity and the density of the ether, on the strength of a certain hypothesis made by Lord Kelvin. In the 11th Edition (1,292) this estimate was repeated, and it was hinted that the hypothesis might be erroneous and the values obtained exceedingly wrong. Everything tends to confirm that conclusion. Strictly speaking the very terms elasticity and density, which are terms applicable to matter, may be inapplicable to the ether without re-definition; if used they must be understood in a formal sense. The properties of the ether are not likely to be expressible in terms of matter; but, as we have no better clue, we must proceed by analogy, and we may apologetically speak of the elasticity and density of the ether as representing things which, if it were matter, would be called by those names. What these terms really

SUMMARY OF PRESENT KNOWLEDGE

express we have not yet fathomed; but, if as is now regarded as very probable, atomic matter is a structure in ether, there is every reason for saying that the ether must in some sense be far denser than any known material substance. The only alternative contention—and it is an important one—is that the density which displays itself as inertia may be due to the organisation responsible for the very existence of matter, and that the unorganised general body of ether does not possess the attribute of inertia. The densest known matter, or matter of highest inertia, is found in some of the stars; the barely visible companion of Sirius having been found, on converging grounds of evidence, to be more than 1,000 times as dense as lead. Unless the above alternative contention turns out true, the density of ether must exceed even that startling amount; indeed there are sound arguments for regarding it as a million times denser. The fundamental substance is not likely to be filmy and unsubstantial.

Recent discoveries have represented the atom of matter as composed of minute electric charges, which fill hardly any of the space inside the atom, so that it is as porous as a solar system. (*See* ATOM.) The great bulk of an atom is occupied only by a few electrons; so that it is by no means impenetrable to particles, which if they fly through it at sufficient speed, may escape being entangled and absorbed. Matter therefore is comparatively a gossamer structure, subsisting in a very substantial medium. An estimate of the substantiality of the medium can be made from its magnetic energies, and it comes out almost incredibly large. If it is right to express it in terms of material properties (which is doubtful) its inertia comes out as of the order of 1,000 tons per cubic millimetre. While as to the elasticity, that is still more enormous, since it is equal to the density multiplied by the square of the velocity of light.

These values are barely conceivable, being so much higher than anything of which we have sensual experience. But still they should be capable of being measured and expressed; so the ether is a physical substance, with properties which can in time be ascertained; and if the estimate above given of the source of the vast energies involved is wrong (as it is sure to be

191

MY PHILOSOPHY.—PART II

inadequately and incompletely worded) subsequent investigation can correct it. Meanwhile we may assume that there is some truth underlying these modes of expression, a truth which we cannot at present formulate any better.

The constants embodying the physical properties of the ether though so huge are not infinite: its properties are finite but very simple and perfect. It is perfectly transparent, it dissipates no energy; otherwise the stars and the spiral nebulæ could not be seen at their gigantic distances across space. There is no friction between matter and ether, otherwise a portion of matter isolated from the rest would cool down, and the planets would not continue for ever in their courses unperturbed. The ether has nothing of what we call in matter viscosity or fluid friction. There is no real heat in the ether, nor any sound. Nothing but one simple type of propagation by waves goes on in free space, and that with a definite unchangeable velocity which is known as the velocity of light, the one fundamental and so to say absolute velocity in the universe.

POSSIBLE STRUCTURE

The question arises as to what the velocity can be due to. The most probable surmise or guess at present is that the ether is a perfectly incompressible continuous fluid, in a state of fine-grained vortex motion, circulating with that same enormous speed. For it has been partly, though as yet incompletely, shown that such a vortex fluid would transmit waves of the same general nature as light waves—*i.e.*, periodic disturbances across the line of propagation—and would transmit them at a rate of the same order of magnitude as the vortex or circulation speed. There remains indeed a question of stability to be safeguarded, but in these days of quanta stability considerations are apt to be deferred. Thus it appears possible that some day an extended hydrodynamics of a perfect fluid will explain all the physical properties of the material universe. *See* Lord Kelvin, "The Vortex Theory of Ether," *Phil. Mag.* (1887) and *Math. and Phys. Papers*, vol. iv. and passim; also G. F. FitzGerald, *Proc. Roy. Dub. Soc.* (1899), or *Collected Papers*, pp. 154, 238, 472.

SUMMARY OF PRESENT KNOWLEDGE

This notion of a structure due to vortex circulation in a perfect fluid may be regarded by some as too material an idea, and it may have to be discarded; but it is the nearest approach that can be suggested to a pictorial image of the etheric constitution. Certainly no *structureless* fluid could transmit actual radiation. And certainly the ether is continuous and without viscosity or any dissipation of energy, and so in many respects is like an ideal fluid. More than that we cannot say, except speculatively, about its constitution.

ETHER AND MATTER

Meanwhile we must assume that the ether has a substantiality and a wave-conveying structure beyond our present clear imaginings, with parts of it modified in an unknown way into electrons and protons; that of these the atoms of matter are built up; and that the whole of material activity consists in the interactions of these minute electric charges, connected as they are by their lines of force and by radiation.

These electric charges, and the aggregates which they have built up are subject to what we experience or recognise as locomotion. The ether itself is stationary. Whether it is really infinite in extent, or whether, though boundless like the surface of a sphere, it is nevertheless finite, are questions which we cannot at present answer. There is no doubt that it extends beyond the farthest visible stellar object, and for all practical purposes is infinite. There is very little doubt that matter is not an alien substance, but is essentially composed of it, being built up of the electrons and protons whose constitution has not yet been ascertained, but which must somehow be constituted of ether, perhaps in some sense analogous to that in which a knot in a piece of string is constructed of string, or a vortex in air is composed of air, or the fibre of a muscle is still essentially flesh.

EINSTEIN'S THEORY

The theory of Relativity has led some people—not many of the leaders of thought—to doubt if the ether can really exist.

MY PHILOSOPHY.—PART II

It may be useful therefore to explain in what way the equations connected with that theory are to be understood physically. Newton expressed the laws binding the planets and suns together in terms of a hypothetical force acting between them, the same kind of force as we experience when a weight is supported above the earth; which force may therefore be taken as a fact of experience. But though the force is a fact, it is not explained: any expression in terms of action at a distance is necessarily incomplete.

Einstein was led by considerations of relativity to formulate a law of gravitation, not in terms of force or of action at a distance, but in terms of something in space, that is, in the ether, which results in a tendency of bodies to approach each other. It might be called a warp in space, or it might be called by other names: the names do not matter. The thing that has to be expressed is that the presence of matter modifies its whole neighbourhood, causing a gravitational potential or virtual stress. And, until we know more about its intimate nature, the action of this modification is best expressed in terms of differential equations which seek to formulate abstractly, without physical hypothesis, the essence of what is really happening. None of the arguments which necessitate the existence of a medium are affected, no name for it need be used, nor need the idea of a medium be introduced, for mathematical purposes. Mathematicians are quite able to work with abstract equations about quantities without physical implications or conceptions, as long as they remain purely mathematicians. They can reduce even geometry to arithmetic.

In a complete expression for the enlarged geometric "interval" between two points, the element of time must be introduced as well as the element of space, because they may be moving points. In other words geometry must be enlarged into kinematics, in order to express activities. The interval or line element between two neighbouring points may be expressed in polar co-ordinates r, θ, ϕ, thus: $ds^2 = -dr^2 - (rd\theta)^2 - (r \sin\theta d\phi)^2 + c^2 dt^2$, a mode of expression devised by Minkowski, an enthusiast for this kind of four-dimensional treatment, where the fundamental etheric velocity c is introduced as a coefficient able to turn time into imaginary space, $icdt$. The emphasis on c,

SUMMARY OF PRESENT KNOWLEDGE

as an absolute geometric constant, is perhaps the most remarkable part of the Einstein-Minkowski conception, as a preparation for the building erected upon it.

But Einstein took a further step, introducing the gravitation potential as something which would modify the motions of matter, and introduced it not only into the element of time (as Newton might have done if he had used that notation) but into the element of radial distance also; so that if the points are in the field of a mass of matter m the Minkowski equation is:—

$$ds^2 = -\gamma^{-1}(dr)^2 - (rd\theta)^2 - (r\sin\theta d\phi) + \gamma c^2 dt^2$$

where $\gamma = 1 - 2P/c^2$, P being the gravitation potential at the place considered; which, if caused by a mass at a distance r, is $P = {}^{km}/_r$, with k as the Newtonian gravitation constant.

Here the coefficient γ occurs twice. If it occurred in the t term only it would be a mode of stating Newton's theory of astronomy, in differential instead of integral fashion; but this γ occurs in the r term also, as a result of the isotropy of the fourfold medium contemplated in this gravitational theory. This equation when elaborated gives, strangely enough, the outstanding progression of the perihelion of Mercury, and it also gives the double deflection for a ray of light passing near the Sun (doubled because the co-efficient γ occurs twice), which has since been confirmed quantitatively by observation of stars near an eclipsed sun. It likewise gives the shift of the spectral lines emanating from any sufficiently massive body, which has now been confirmed beyond the reach of reasonable controversy by observations on light coming from the companion of Sirius, which Eddington has astonishingly proved to be by far the most compact and densest material body at present known to science, so that it is characterized by an excessively high gravitational potential.

The beauty of these results is overwhelming; but the idea that any mathematical scheme is more than a powerful method of exploration, and that a universe can be thus constructed in which physical explanations can be dispensed with, involves too simple and anthropomorphic a view of nature. The things calculated, and the things observed, however brilliantly accordant, cannot exhaust reality; an explanation is bound to be sought, and ultimately attained, in terms of the partially recognised but largely unexplored properties of the entity which fills space.

LOCOMOTION OF MATTER

The locomotion of matter is perhaps the commonest fact of experience, and it seems strange that it should be in need of explanation. But since an atom of matter is composed of electric charges, the locomotion of those charges has to be considered more in detail. An electric charge in motion constitutes an electric current, and the path of every electric current is surrounded by rings of magnetic force. This magnetic field confers inertia or momentum upon the moving charge; so that a mechanical impulse is necessary to start it moving; and also to stop the motion. If not stopped it will continue to move uniformly in a straight line until it encounters some deflecting or retarding agency.

But though locomotion can thus be stated and worked out electromagnetically, that cannot be regarded as an ultimate explanation of so familiar and apparently simple a thing. Moving matter is known to have kinetic energy; and the familiar expression $\frac{1}{2}mv^2$ is the type of its measure. But when we come to analyse this expression there are difficulties about it, which hardly need elaborate theory to bring out and emphasise. For when we try to specify the velocity of a body, in order to calculate its energy, we find it difficult to say what that velocity really is: we can only specify it with reference to something else, commonly with reference to the earth. But the earth itself is moving. Hence $\frac{1}{2}mv^2$ does not give the absolute energy, but only the energy relative to the earth or other frame of reference, as Newton implicitly recognised. What the velocity of a body is in space we have no means at present of ascertaining, having no universal standard of reference; and accordingly the usual expressions, though practically useful, are by no means ultimately satisfactory. Nor can a statement in terms of electromagnetism, be considered as ultimate.

The fact is that locomotion does not seem to be a property of the ether; ether appears to be affected by one speed and one speed only, namely, what we may imagine to be the speed of its internal

SUMMARY OF PRESENT KNOWLEDGE

circulation and are familiar with as the velocity of light. Yet a modified particle of ether, like an electron, can move from one place to another. The analogy of a loose knot slipping along a string may be helpful.

An electron even at rest has intrinsic energy, *viz.*, its electrostatic energy of constitution, which can be expressed in various ways, and which, when expressed in terms of mass and speed, is m_0c^2, m_0 being its inertia at rest. Its static energy is thus expressible as equivalent to that of a particle of certain mass m_0 or $2m_0$, moving with the speed c, the speed of light. To assist ideas, it might be thought of as a spinning motion; at any rate not locomotion.

When the particle is moved, the natural idea would be that this velocity c is increased, or that some addition is made to it. But according to the doctrine of relativity that is impossible: the velocity c is constant. The thing that changes is not c, but m. And the energy of a moving body is m_1c^2, where m_1 is greater than m_0. As the speed of motion increases, m_1 increases too; until at high speeds it grows fast, and, as the speed of light is approached, tends to become infinite. The factor, or ratio between m_1 and m_0, is $c/\sqrt{(c^2-v^2)}$. So when an identified portion of ether is in locomotion, it is not the fundamental speed that is changed, but the amount of modified ether, or modification of ether associated with that identified moving portion. And what we observe as the kinetic energy of a body is really $(m_1-m_0)c^2$ or c^2dm. This is what we have hitherto recognized and called $\frac{1}{2}mv^2$, an expression which is only relative, and moreover is not exactly applicable to great velocities, such as we encounter in vacuum tubes and in radioactivity generally.

SHORT SUMMARY OF PRESENT KNOWLEDGE

To sum up our present position in more compact form, in order to put on record what may perhaps excite the interest or else the derision of posterity:—Assuming the Ether to be in some sense a substance, that is real and substantial, a genuine entity and not the mere emptiness which it superficially appears to be, the things that are known about it are these:—

(1) It is absolutely transparent and undispersive. In other words it quenches no light but transmits it undiminished in total

197

intensity, though diluted by spreading, to and from the greatest distances known in astronomy. Moreover it transmits every kind of radiation at the same pace, whatever the wave-length, except in so far as it is interfered with by electricity or matter.

(2) It is absolutely devoid of viscosity. In other words it allows the motion of matter through it without any friction; it dissipates no energy and generates no heat. A serious attempt made at Liverpool (University College) from 1890 to 1897 to detect a mechanical grip or cling between ether and rapidly moving matter, failed (as was more than half expected) to find any convective effect, even when the moving matter was strongly electrified or magnetised. (*Phil. Trans. Roy. Soc.*, 1893 and 1897.)

(3) Ether is the sole vehicle of radiation, that is of transverse disturbances periodic in space and time travelling at a definite and immense speed without any obvious destination, but it neither emits nor absorbs them. In other words it generates neither heat nor light; but it can receive these forms of energy from matter, and can convey and deliver them to other matter at a distance. Our sensory instrument, the eye, has long familiarised mankind with various practical aspects of this wonderful phenomenon.

(4) An electric field is another form of energy existing in the ether. For this we have no sense organ, and therefore know less about it, but its lines of force appear to be of the nature of strain. And probably the ether is the seat of all strain or potential energy. An electric field (like radiation) can only be originated by matter: its lines of force never terminate in ether, but they pass through ether along their whole extent. Insulating matter only modifies the lines, but conductors stop them.

(5) Another etheric form of energy is a magnetic field, which is certainly different from an electric field though somewhat similar. Magnetic lines of force are closed curves, and seem more intimately connected with the ether than with matter. But they interact with matter, and have thus displayed their existence by consequent attractions and repulsions.

(6) Electric and magnetic fields interact also with each other in free space, and thereby constitute radiation, in accordance

SUMMARY OF PRESENT KNOWLEDGE

with the Poynting formula that the flux of energy at every place is their vector product.

(7) Chemical affinity between atoms of matter is undoubtedly due to electric or magnetic attraction or both. And cohesion may be attributed to a residual chemical affinity between molecules. Thus the ether is indirectly responsible as a vehicle for all physical and chemical activity, and no one who believes in the ether has any doubt that it is responsible also for whatever is represented by the word "gravitation." What other functions this universal medium may be found to possess, and whether life and mind can be in any way associated with those functions, must be left to posterity to find out. Our serious surmise is that they are so associated, in a primary sense, and are temporarily manifested by secondary association with matter.

STEPS TOWARD FURTHER KNOWLEDGE

Beyond definite knowledge, other guesses and working hypotheses have been made concerning the ether on the assumption that its properties can be partially expressed in terms of more or less familiar ideas. The property of inertia, so fundamentally possessed by matter, is doubtfully applicable to the ether. Even if matter turns out to be really modified ether, as many are beginning to expect, it is a question whether inertia arises as a result of the modification, or whether it is a property of the primitive substance which, by the materialisation, is individualised, localised and made manifest. If inertia can rightly be predicated of the ether itself, its value per unit bulk must be enormously greater than is exhibited by any kind of matter; for matter by its very constitution is certainly excessively porous, consisting as it does of minute particles far apart from each other in proportion to their size, whereas the ether must be as continuous as space itself. A molecular structure for the ether is not to be thought of, for its whole value as an explanation of facts depends on its continuity: separate particles with interspaces are appropriate to matter, but the whole problem of action at a distance would remain unsolved unless the particles are united into a coherent whole

199

by something which has no gaps, and is responsible for cohesion, elastic rigidity and other properties of solids.

The fundamental units of measurement, the centimetre, gramme and second, have direct relation to matter, and it is doubtful if they are applicable to the continuous ether at all. If they are, then there are grounds for maintaining that the inertia of unit volume of ether is represented by a number of the order 10^{12}; while, since it certainly transmits the polarisable and therefore transverse vibration of light, it must on that view have a quasi-rigidity comparable to the number 10^{33}.

This elastic quasi-rigidity can be attributed to a continuous perfect fluid provided, and only provided, it is in an excessively rapid and fine-grained state of vortex motion; and Lord Kelvin showed that such a rotational or turbulent fluid could transmit transverse waves at a speed of the same order as the circulation velocity. This velocity c is now regarded as one of the unalterable constants of nature: it is the one definite measurement which has been made concerning the ether of space, and of itself is sufficient to show that space empty of matter is endowed with finite and measurable physical properties. We can measure the speed of light by aid of matter, because matter generates, absorbs, reflects, and otherwise interferes with it: we observe electricity and magnetism and every other manifestation of the ether by aid of matter; but unfortunately all the properties of ether itself, apart from matter, have hitherto proved completely elusive. None of our apparatus grips or gives us a foothold; so that some physicists claim that pragmatically the ether is a gratuitous hypothesis and need not really exist. It is quite true that physical calculations and discoveries can proceed without explicit reference to the ether, but when we come to philosophise and try to formulate the facts physically, it is clear that space must be endowed with physical properties and is therefore entitled to something more than a merely geometrical name. These properties are equally real inside matter, and radiation is everywhere conveyed by space: transparent material does not really convey light, it only allows the passage and reduces the speed.

So much for a transparent body, which must be an insulator because the electrons are tightly attached and not free to move.

SUMMARY OF PRESENT KNOWLEDGE

On the other hand, when the electronic constituents of matter are loose, not anchored to something heavier than themselves, the substance becomes a metallic conductor, and as such must be mainly opaque. A conducting film, or rarefied electric region, if it can transmit radiation at all can only do so in a peculiar manner. In an electrified region waves do not travel as in free space. Different wave lengths begin to be treated differently, for their speed is a function of wave length. An expression for their speed in that case is

$$v = \sqrt{(c^2 + b^2\lambda^2)}$$

where b^2 is proportional to the electrical concentration. It turns out that b is the smallest frequency which such waves can have under the given conditions. Strangely enough the energy of the radiation is apt to lag a little behind these curious waves, for it travels at a speed called the group velocity c^2/v; and this may be slow when b^2 is big. The amount of energy is proportional to the frequency of vibration.

INTERACTION OF ETHER AND MATTER

A part of space occupied by matter or electrical particles transmits radiation in a peculiar way. Treated in a statistical or average fashion, matter in which electric constituents are firmly attached to the atoms—so that it insulates when solid, and conducts chemically when liquid—has a refractive index μ which reduces the apparent velocity of light to c/μ; a simple consequence of wave theory which Foucault definitely verified; though the full explanation of such a reduced velocity is not simple. Maxwell's view of the dielectric constant, or specific inductive capacity of insulators, is that it must be nearly the same as μ^2. Transparent matter thus seems to load or increase the effective density of the ether by the amount $\mu^2 - 1$, so that what is sometimes spoken of as bound ether—the portion appearing to cling to matter and move with it—is in such a substance $\dfrac{\mu^2 - 1}{\mu^2}$ of the whole: as Fresnel surmised and Fizeau

MY PHILOSOPHY.—PART II

experimentally verified in 1851 by a successful experiment on the speed of light in moving water.

It must be admitted that this is only a superficial or tentative way of regarding the still partially unexplained reaction between matter and ether; for it must be understood that statistical or average forms of statement are never completely and finally satisfactory; they fill a gap in our knowledge for the time being, and are true as far as they go. The Lorentz transformation, used by relativists, arrives at the same result without philosophising about it or explaining it.

QUESTION OF REVERSIBILITY

Every star is emitting energy at the expense of its own material, so that matter is gradually turning into radiation and passing into an unlocalised form in the ether. It may be said that, without the restriction of the quantum, whereby only whole units of energy can be radiated, all the energy of matter would pass into the ether and become radiation. A good deal does. The question naturally arises whether this process is reversible or not; *i.e.*, whether radiation can under any condition generate, in return, the fundamental ingredients of which matter is composed. This discovery has not yet been made. What we know of is that the jump of an electron generates radiation, of a frequency proportional to the energy of the jump; and that this same radiation, whenever absorbed, can cause another electron elsewhere to jump with the same energy. Hence the idea looks hopeful that a reversible process may be involved generally, in the interchange of energy between ether and matter, not only in this ordinary electronic laboratory case, but in the more violent clashes in the stars, where matter appears to be destroyed. May it not perhaps in some distant region be reconstituted, with a consequent great gain of gravitational potential energy, so as to render the cosmos permanent, and reduce the useful law of dissipation of energy to comparative insignificance?

CONCLUDING REMARKS

We have seen that when we try to look at even so apparently simple a thing as locomotion, absolutely, we have to admit

SUMMARY OF PRESENT KNOWLEDGE

that varying speed means varying amounts of ether-modification in the identified portion of matter we are attending to; for we can only express absolute energy in terms of an ether constant c, which at first sight would appear to have nothing to do with it. The same constant enters into the composition of velocities. It is as if the normal constitutional etheric circulation trended or drifted in one direction, so as to constitute perceptible or available energy, much as the energy of a river or a gale of wind is a directed fraction of random molecular motions.

The same idea may be expressed magnetically by calling attention to the magnetic field surrounding a moving charge. At high speeds the magnetic field is strong; more substance is involved in it: and the additional spin (if that is the right term, for magnetism is usually thought of as a kind of spin) accounts for the additional energy. Why it should appeal to us as locomotion, and what the real meaning of locomotion is, are not so clear. This is only an illustration of the difficulty we experience when we come to probe the simplest thing to its depths. We have grown accustomed to certain aspects of the universe given us by our senses, but we do not fundamentally understand them. And when we come to probe the meaning of things deeply enough, we find ourselves up against difficulties of conception, toward the elucidation of which our senses give hardly any aid. What we are used to is mechanical movement; but the effort to explain things ultimately in that way is not easy, and may turn out to be not possible.

Meanwhile we take refuge in expressing these things in terms of electricity and magnetism; which is a step toward an explanation, and is useful in bringing out the difficulties which underlie every ultimate and absolute statement. The attempted absolute expression for static electric energy, mc^2, with the inertia m as the only variable, is a legitimate mathematical expression of actual facts. But the real meaning of c is, at present, a hypothesis: and what the real meaning of m is, must be regarded as still less known. Both these factors must have reference to the ether, and until we know more about the constitution of the ether we must be content to remain in a condition of provisional ignorance. We are led to regard the material universe as a substantial reality in various stages or varieties of internal activity. We may try to think of this activity as akin to a fine-grained vortex circulation in a continuous, incompressible, perfect fluid: beyond that we cannot at present go; nor are we clear about the exact meaning of these terms when applied to a medium of unknown constitution. When we understand the real and ultimate nature of electricity and magnetism we may hope to proceed further. Till then we must be content with proximate explanations and await the gradual illumination of further experience.

PART THREE
THE INTRODUCTION OF LIFE AND MIND

CHAPTER XVIII

THE INTERACTION OF THE PSYCHICAL WITH THE PHYSICAL

"I have assumed that man is an organism informed or possessed by a soul. This view obviously involves the hypothesis that we are living a life in two worlds at once: a planetary life in this material world, to which the organism is intended to react; and also a cosmic life in that spiritual or metetherial world which is the native environment of the soul."

F. W. H. Myers.

The universe as a whole contains matter and motion, but it contains more. Experience shows that it contains also mind and spirit: and these diverse aspects or regions of existence interact with each other, so that the phenomena of animation and consciousness result. Some matter undoubtedly appears animated: and if we limit our studies to that which is not animate, and attend only to the inorganic realm, we may be excluding a vitally important element, without which a philosophic understanding of the universe is impossible.

A physicist feels the difficulty directly he begins to think about origins. He is accustomed to take the material world, with all that it contains, as a working mechanism, into the origin and meaning of which he does not enquire. Or if he does so, he is trying to exercise the faculties of a philosopher, rather than of a pure physicist. Considered narrowly a physicist is supposed to be exercised chiefly with measurements,— with metrical determinations or pointer-readings: but even as a man of science he is not satisfied with that limitation; and as a human being he cannot ignore at all times the mystery of existence and the problems presented by live things. He is beginning to suspect that even the tools with which he works, the abstractions he has dissected out from his observations, such as space and time and matter, have their roots deep down in something which he has not explored, but in which the philosophic solution of the whole must be sought, if it is ever to be found. He perceives that the sciences of chemistry and physics, of biology and psychology, are not separate sciences, independent of one another, but are really parts of a unified whole; and that this whole must be taken into consideration in attacking any ultimate problem, and must be comprehended before a solution can be found, that is to say before the universe becomes really intelligible.

Every scientific man knows that his special object of study is but a small part of a gigantic whole, and that to attain genuine comprehension he must enlarge the scope of his theories, and make some effort towards a more comprehensive solution. The universe of matter seems complete up to a point. The laws of its inter-actions with the parts of itself can be investigated, and form a great body of knowledge, which within limits is satisfactory. He may limit his studies to the action of the mechanism, and profitably expend a life work in that great and important study. But he knows by his own experience that there are other things which, as a physical investigator, he purposely excludes from attention, such as life and mind and consciousness; though these clamour for notice when he is in a philo-sophic mood, and may contain the key to the whole performance. All that physical science tries to study in any process is the "how"; it has nothing to do with purpose. The "why," if it can be understood at all, must be a problem of another enquiry: the purpose of the Universe may not be plain to a metaphysician, or even to a Theologian. It may lie in a region beyond present human intelligence. We need not ask questions like "why" constantly or prematurely, but before we dogmatise philosophically we should emancipate ourselves from the limitations of any partial view of existence as a whole: people who are still immersed in limitations are not entitled to the honourable and comprehensive title of Rationalist or Freethinker on which they pride themselves.

Every now and then a physicist has become aware, by inference based on experience and observation, that there is a spiritual world, as well as a material one, that the two constantly interact, and produce effects by means which for the most part he deliberately ignores. The first hint of this discovery is supplied by the activities of his own body. He finds that the ideas and conceptions associated with mechanism are insufficient to explain even the stretching out of his finger, or any other spontaneous movement, directed to a specific purpose. He finds himself endowed with intelligence, emotion, and will. He uses these in his ordinary investigations, though the use of them is so instinctive that he is often not aware of what he is doing; but when he attends, he cannot really explain the sensations which the material world produce in his mind. He learns, for instance, that sound is a vibration of the air; but the sensation of sound is quite different from any measured vibration. He finds that the tones of a voice can be expressed as different frequencies; but the impression made upon him by those tones is something quite different from the rate of a mechanical disturbance; so that a certain sequence of vibrations, called words, can not only appeal to his intelligence, but can stir his emotions to their depths. Again, when he responds, as he does through his nerves and muscles, he is expressing something far more intelligent and full of feeling than could be expected of any mechanical movement, which is all that physically he is able to accomplish. And so it is with his other senses. The touch of an object may be interpreted with a rich significance that is not explicable by any mere contact, and when he opens his eyes in some surroundings he is impressed with a vivid sense of colour, and it may be of beauty, so that it can become for him a part of himself never to be forgotten,—a scene that cannot adequately be expressed in terms of different frequencies of vibration; though these are all that physically he receives.

Sensation in fact is not only the tool by which he explores the external world, the means whereby all his experiments are conducted: it is something which would not exist without his mind. The mind interprets the physical stimulus in a certain way; and the interpretation, though instinctive and habitual, cannot be reduced to, or accounted for by, any form of mechanism. He is thus already and constantly in touch with another aspect of reality; although in its interpretation, for the most part, instinct takes the places of consciousness, so that the remarkable peculiarities of the mental impression are usually ignored. There is, so to speak, another world, interacting with the world of matter; and of this mental or spiritual world he is aware so directly and so constantly that it has ceased to be surprising or even noticeable and is taken as a matter of course.

The same thing is noticed, or can be noticed, in the most elementary processes of physiology, such as what is spoken of as metabolism. A live creature, though it may be only an amoeba, has the power of assimilating other material into itself, converting it into its own organism, and thus growing and obtaining energy for its movements, so long as the vital spark remains. The food does not determine the shape of the organism; that is determined by the organism which assimilates the food. If it were not so, it might truly be said that "one is what one eats," in accordance with the German proverb. The converting and assimilating property is found not only in the organism as a whole, but in every part, and apparently in every fraction of each part. The appropriate atoms in the food are selected and manipulated, doubtless according to physical laws, so that the structure they produce at any given place is a bone or a muscle or a hair or a feather, a nail or a claw, or even the proper portion of an eye or an ear. The determining action is localised, and is a result of the animating principle in the organism interpenetrating and existing throughout its different parts, yet all controlled by a unity of design. The cells of a feather, for instance, or a hair, are not deposited at random. The place where they are deposited determines the subsequent colour and appearance of the whole animal, but is itself determined in accordance with some general scheme; so that the animal as a whole shall be striped or spotted in accordance with a general design; or even so that the colour may be varied according to the circumstances in which the animal finds itself, so that the creature can correspond with its own background, and thus escape the attention of other so-called hostile organisms, who, if they were aware of its presence, might use its body prematurely for food. This contribution to the life history of some other organism will doubtless be its ultimate fate; but meanwhile it has the opportunity of reproducing its species, and so continuing the existence of its own particular variety of organism, in accordance with the laws of heredity. In some predatory animals the concealment thus provided serves not merely as a protection but enables them more successfully to stalk their prey.

MY PHILOSOPHY.—PART III

In all this there is evidently something more than meets the eye. I have only given a sample of a whole multitude of phenomena studied in the great subject of biology, which are more or less familiar to everybody, so familiar that we have to some extent lost the sense of wonder, and are apt to take them as merely examples of the powers and properties of animated mechanism. Many biologists quite properly attend to the material processes, and seek to ascertain the chemical and physical concomitants of the strange powers of adaptation displayed by animated matter. But in the last resort they will find that the clue to adaptation is to be found in the mystery called animation, that is to say the interaction of a mind or anima, or of a something unknown called life, which interacts with and guides the material processes to a destined end.

Sometimes it is true the process may be guided wrongly, or rather deflected from its natural path, by the intervention of some other animated organism, a parasite or microbe of some kind, or by some foreign ingredient, which may be called a poison, that deflects the atoms from their normal course, and causes them to make cells either out of place, or too lavishly, or too meagrely, and thus give rise to some of the various phenomena we call disease. This is part of the interaction of organisms on one another, and may be susceptible of physiological, or in some cases of chemical and physical, explanation. But still at the root there is always the animation to be considered, and that is not explicable in terms of chemistry and physics. Biologists find that the matter of the organism differs in no respect from other matter, except perhaps in complexity of atomic constitution. Food for the most part consists of material atoms that have been put together into complex molecules by the agency of antecedent life. Only a few organisms are able to

PSYCHICS AND PHYSICS

assimilate the crudely simple molecules of the inorganic world, and those mainly, as in the vegetable kingdom, by aid of the etheric variety of energy received from solar radiation. Still it is only when the plant is alive that it is able to do this. The guiding and controlling entity we call life must still be held responsible. Life need not have risen to the status that we call mind or consciousness, but still it is the rudiment, and exercises unconsciously some of the powers which we observe in our own conscious and purposed actions.

The specific activity of each part is not due to a system of central control worked by a nervous system; for it has been found that the right structure is formed by a portion of tissue isolated from the rest *in vitro*. So testifies the experimental work of Dr. H. B. Fell, head of the Strangeways Laboratory, Cambridge.

Life does not belong to the material world; though it may be a function of the etheric world. All we know of it is the animation or vitality which it produces in certain kinds of matter. Its properties are displayed by the behaviour of the organisms which it animates. In itself life is more akin, and properly belongs, to the spiritual world; and so we become aware, in a sense, of the interaction of the spiritual or at least the etheric world with the material at every grade of existence, wherever matter is animated by either vegetable or animal life. We shall never find the clue to the behaviour of organisms so long as we attend to matter alone. Organic matter is just like any other matter: it obeys the laws of physics and chemistry perfectly. It has not any kind of spontaneity, it is perfectly inert; it obeys the laws of motion, it takes the path of least resistance, and when set in motion, has not even the power of stopping. Of itself it can do nothing. Yet it is through the behaviour of this inert matter that we have to investigate the properties of what is acting on it. We have to use it as an index or demonstration of something which otherwise would be unperceived. Don't let us make the mistake of saying that the organism is *nothing but* the material manifestation.

It is the same with many other agencies. We have no sense for an electric current except for the effect it produces on matter: we can detect the heat it produces, or the chemical action, or what we call the magnetic effect; otherwise it would escape our notice. All that we can see is the deflection of a compass-needle; we take that as the sign or demonstration of an electric current: we do not use the phrase "nothing but" in that case. Whatever an electric current may be, it is certainly not a galvanometer deflection.

So it is with the other physical agencies; we only know of light when it acts on matter (we only see a beam of light when it illuminates the dust of the air); we only know of magnetism by the aid of a substance like iron. We have no direct apprehension of a multitude of things, which nevertheless we explore by the material instruments which they affect. The agencies themselves are in the ether. So there is nothing peculiar in that respect about the agency of life. We are bound to explore it by means of the behaviour of animated matter. And as for the agency of spirit, the means by which we can explore that is through its interaction with our own minds. Mind certainly is not physical, but our manner of exploration is largely concerned with the way it operates on the material organisms we see around us, and with inferences based on their behaviour.

CHAPTER XIX

LIFE AND MECHANISM

"There is a certain congruity between our logical activities and the world to which we apply them."

PROFESSOR McDOUGALL.

The doctrine of mechanism and physical causation is very seldom sufficient to give a complete account of any ordinary occurrence involving the action of live things. It may be true as far as it goes, but it does not go all the way. Sooner or later we find we have to appeal to some psychic fact, some action of what we may call a mind, even though it be a mind in incipient form. The first and most obvious criterion for some activity not within the scope of physical science is spontaneity of movement, so that our instinctive remark about the spasmodic action of, say, a jumping-bean, is that "it behaves as if it were alive." Clockwork mechanism may be made to simulate vitality, and might deceive a child for a short time, but it is always sooner or later perceived to be an imitation, the mechanism of it is completely intelligible and may be readily understood. But sometimes the thing really is alive, as for instance in the case of a jumping-bean. Inside the hollow space is a maggot or other creature, which coils itself up, and suddenly liberates itself, so that it jumps against the ceiling of the cavity, and makes the whole thing leave the ground for an instant. As soon as it is understood that the action is that of a live thing, no further explanation is thought necessary, and physics

215

loses interest in it as soon as it has explained that the action is consistent with the laws of mechanics, given the unexplained spontaneity. If any live thing shows a behaviour which is difficult to explain mechanically, then a physicist might be interested, as for instance the way in which a cat, if dropped from a small height, manages to alight on its paws. This was seriously examined by Clerk Maxwell, and shown to be consistent with the law of conservation of angular momentum, though at first the rotation of a body in mid-air did not seem easily explicable, unless it possessed an initial amount of rotation. The wriggling of a falling animal may instinctively achieve a result which is at first puzzling; yet in the end a complete explanation may be, and always is, forthcoming, though the instinct responsible for the bodily movements may require some further and more psychological study.

I should suppose the same kind of thing to be true of the movements of a so-called "medium," who in trance may be able to produce physical phenomena of an unusual kind. I should always look for the physical concomitants of such a performance, and assume that they were explicable by ordinary laws. But there might be some intelligence shown by the activity which could not be thus explained, and which would involve some psychical addition. Considering that the medium is alive, it is not surprising that such an element must be appealed to, though the identity of the intelligence concerned may raise questions not so easily settled.

But without entering upon what may be regarded as a questionable set of phenomena, it is quite familiar that a psychic element enters into our everyday experience; and often we are aware that there must have been some plan or design or purpose associated with some simple observation, although we may be unable to specify what it is in any given instance.

LIFE AND MECHANISM

If it be true that we have to postulate the inter-action of an ether to explain the movement and the behaviour of every particle of matter, it may also be true that we have to postulate the agency of that same ether before we can understand the way in which a psychic element operates on the material so as to bring about a purposed arrangement. For if ever the psychic and the material interact, there must, one would think, be some physical agency which connects the two, so as to render the action possible; that is to say, there must be some physical mechanism which can transmute a thought into an act. It is by such transmutation into material activity that everything is accomplished. Trans-formation of potential energy into kinetic or *vice versa* is the accompaniment of every activity, and I argue that transmutation from the psychical to the physical is the accompaniment of every purposed action carried out in accordance with an intelligent design, whether the intelligence be of the lower grade of instinct, or whether it be of the more dignified kind which we associate with consciousness.

Why do I bring the ether into such an ordinary everyday process? The reason is because physics has taught us that inert matter can only be operated upon through the ether, and is never otherwise interfered with and can do nothing of itself; as I have already frequently said; and because we find by our everyday experience that intelligence does produce results in matter, and therefore must have been able to operate on the ether in some way to which we are accustomed, but which we do not yet understand. The physio-logical mechanism by which an act is achieved is complete up to a point, but it never explains the purpose which determined the performance of that act at a given time and place for the attainment of a desired result. Instinct causes a spider to spin a web, a bee to construct

217

a honeycomb and fill it with feeding matter collected from plants as a provision for a time when food is less plentiful; instinct causes a bird to build a nest for the sake of future offspring; and brings about many other structures of beauty and design, such as biologists tell us of. A higher kind of conscious intelligence enables man to design and build a bridge or a house or a cathedral, or to produce a poem or a work of art, for reasons which in some cases he might find it difficult to specify. He works completely in accordance with the laws of physics, but the laws of physics alone would never account for his performance. The physical world is insufficient, the psychic or spiritual world must always be appealed to. The interaction of such a world is displayed in its lowest grade by the totally unconscious agency of mere vitality, such as is exhibited by vegetation, where the energy from the sun is utilised to produce foliage and flowers and reproductive organs all perfectly adapted to their purpose when not interfered with by untoward agencies; and these may be rightly taken as an indication of some planning and design not in the least incorporated in the physical and chemical mechanism concerned.

If we consider the lilies of the field in connexion with the other phenomena that we have been dealing with, we shall realise that their beauty and adaptation represents some aspect of the universe higher than can be attributed to the properties of matter, higher indeed than anything to be experienced from the interaction of ether and matter, and that requires a planned scheme of guidance whereby etheric energy shall be so influenced as to contribute to and make possible the observed result. The chemical details of the process can be studied by Baly and other chemists, and the elimination of oxygen as a waste product— deleterious if left in the tissues but a health-giving

LIFE AND MECHANISM

and necessary ingredient for animal life when extruded —must be taken as part of the beneficent design which humanity is now becoming able to recognise throughout both the animate and the inanimate worlds.

Indeed we find that the etheric energy derived from the sun is not only conspicuously utilised by vegetation, so as to contribute food and other elements necessary for animal life, but the rays of the sun are being found to have a directly beneficent influence on animal life itself. The vitamins to which so much attention has been recently directed by the President of the Royal Society are I believe more influenced by etheric activity than we are yet fully aware. Bio-chemists have traced their activity in producing Vitamin D, but their influence is suspected in other vitamins also; and without them the chemical ingredients alone, such as could artificially be put together, would be ineffective. This is a matter that requires still further study. The influence of irradiation is being advocated and experimented upon by Dr. Chalmers-Watson of Edinburgh, and I hope by many other bio-physicists. It is known that the chemical ingredients of sea-water artificially put together are insufficient to maintain marine life in a flourishing state. Some other non-chemical but presumably physical ingredient is required, too. It is contained in natural sea-water, and whence can it be derived except from that hitherto ignored physical element the ether, whose radiation activity is known to be responsible for all other energy, and probably for much else of which we are only beginning to be aware? I call upon bio-physicists to supplement the work of the bio-chemists, and to show that the etheric activity of radiation must be appealed to both for the maintenance of health and for the restriction of those subterranean phenomena that we call disease. This must be done before the whole assemblage of biological phenomena are properly understood, and their interacting agencies thoroughly correlated, so as to interpret the great variety of healthy vitality which we find on this singularly well adapted planet.

Thereafter when our existence continues amid other surroundings we shall not be in ordinary association with matter as we are here, and therefore shall not appear to the senses of our fellows left behind on the earth. We shall have a body or mode of manifestation suited to our new surroundings and shall be fully perceptible to our fellows in like case. My hypothesis is that this body or more refined mode of manifestation will be composed of ether, and may be properly spoken of as an etheric body, or what St. Paul called a spiritual body.

CHAPTER XX

A PSYCHICAL FUNCTION SUGGESTED FOR THE ETHER OF SPACE

"Whether this vast homogeneous expanse of isotropic matter is fitted not only to be a medium of physical interaction between distant bodies, and to fulfil other physical functions of which, perhaps, we have as yet no conception, but also, as the authors of the 'Unseen Universe' seem to suggest, to constitute the material organism of beings exercising functions of life and mind as high or higher than ours are at present, is a question far transcending the limits of physical speculation."

From article on "Ether," by CLERK MAXWELL in *Encyclopædia Britannica*, Ninth Edition.

By "spiritual body" St. Paul did not mean one made of spirit, but one that served the needs of the spirit; just as by "psychical body" he did not mean one made of "psyche," but meant the material body which served the psychical or mental need. The psychical body is made of matter and used by soul. So also the pneumatical body is one used by spirit and made of X. My hypothesis is that partially and approximately X = Ether.

This view is not materialistic in the ordinary sense. It does not even claim any direct association of mind with matter: it claims association with an intermediate substance. The etheric body is intermediate between matter and spirit, for it seems probable that spirit requires some kind of physical vehicle for its manifestation. My hypothesis is that spirit primarily inhabits the ether, uses it, and acts on it: and that occasionally this operated on ether is able to act upon matter. Thus through the intervention of ether, spirit can be brought into relation with matter, indirectly: and the intervening mechanism (if it can be called mechanism) is the etheric or spiritual body. The argument is that spirit cannot act on matter directly: as a matter of fact we never act on matter directly, even now, but only through an etheric link; our primary action is always on ether.

Some link is needed between spirit and matter; and the ether offers itself as a substantial, physical, but at the same time strictly immaterial, link. We have still to explain how any portion of ether can be animated and acted upon by something not belonging to the physical universe; and it may seem as if the intervention of another thing, neither matter nor spirit but still physical, is no particular help and does not get over the difficulty. Certainly some difficulty remains; but it is pushed on a step, and the only question is whether that step is in the direction of truth. All our explanations only push the solution of a problem one step forward: they never remove or evade the fundamental mystery. The discovery that glands of internal secretion affect bodily development through chemical substances called hormones so as to exert an influence on character, is a striking advance which at present is attracting attention, but it would be a mistake to consider that as an ultimate explanation; the secretion and function of such substances raise a number of new questions. Meanwhile I venture to promulgate a view which goes beyond and includes all physiological discoveries.

What we have learnt physically is that the ether can act on matter through electric and magnetic properties: we also know that mind can somehow act on matter, though probably indirectly. Our assumption is that we possess an ether body or animated structure of modified ether here and now, that life or mind is closely in touch with the ether body, and that through its action on this at present imperceptible body it is able to exert an action on the familiar material body. To assume that mind acts on ether and that ether acts on matter, is I hope an assumption in the direction of truth: and it appears to be justified by psychical facts, which show that the action of mind can be independent of matter, though probably not independent of everything substantial or physical.

Such a view would tend to harmonise the instinct of the idealists with the instinct of the materialists. The strength of the materialistic position is that *some* physical agency is necessary: hitherto supporters have sought that agent in matter, and by most philosophers are now considered mistaken. If they can be accorded the use of another substantial reality,—one which they have hitherto ignored, and which does not appeal directly to our animal senses,—they may recognise it as an advance in their direction: and at the same time it is an advance which may legitimately be made by idealists, who must have felt the difficulty of completing their scheme without some insensible intermediary. The view that the intermediary is a substantial reality, about which clear conceptions can gradually be formed, is not a view to which it seems that we have a right to object. Physics is already learning much, and may hope gradually to learn more, of the interaction between ether and matter: and it would be left to psychologists to learn more (if they can) about the connexion between mind and ether. They must begin by grasping the fact that the ether can be animated; the connexion between mind and ether must have laws which can gradually be explored,—though at present we have hardly a clue.

For a long time we had no clue as to the connexion between ether and matter; but the connexion was there all the time, it only remained to be discovered. Faraday and Clerk Maxwell began the discovery of the link between ether and matter: it may be years before the other link, the link between spirit and ether, begins to be discovered. But it will be a step in advance if attention can be concentrated on the problem in that form: at present we can only have speculation. The speculation was probably begun in some vague form here and there by an inspired genius long ago; but it was set forth more clearly, at least as a legitimate speculation, by Professors Balfour Stewart, of Owen's College, and Peter Guthrie Tait, of the University of Edinburgh, in their book of last century called *The Unseen Universe*. And to-day the increasing number of facts which have come to light tend on the whole to support their speculation. Speculation is not science, but it is a necessary preliminary; and it may tend to become more definite, and so gradually more scientific, as time goes on.

At present the only link between mind and matter generally recognised is that the mind has some connexion with the cells of the brain: but the phenomenon of ectoplasm (when that becomes more thoroughly established and more generally recognised) seems to show that brain cells are not the only form of protoplasm which can be manipulated by intelligence. And this in itself may be held to constitute, in due time, a remarkable discovery, and suggests another region in which a clue to the connexion may be sought. For it may be easier to investigate the psychical properties and behaviour of protoplasm when that organised material substance does not form part of a completely formed and customary organism, to the complicated actions of which we have grown too accustomed. Statements about what can be done with ectoplasm are very similar to our familiar experience of what can be done inside

A PSYCHIC FUNCTION FOR THE ETHER

our own bodies, where material is assimilated and formed into certain shapes or organs, and used to produce intelligent vibrations, direct speech, and the like. Also we can remember that full-blown materialisation of another spirit can occur inside the body of a female, through the agency of a placenta which is manufactured for the purpose. It now appears that several things which our own sub-conscious intelligence is able to do with the material supplied to our bodies are capable of being performed, in a less permanent and less continuous manner, by other intelligences, which make use of what appears to be some already elaborated and extruded protoplasmic material for the purpose.

The facts seem incredible because they are new: many well known facts would be equally incredible if we were not accustomed to them. The formation of specifically shaped bodies out of any kind of food, and the use of those bodies for the reception and transmission of intelligence, is as great a mystery as any of those which are now asserted and controverted and disbelieved in. These last phenomena have to make good: they have to show themselves worthy of critical examination. When they have arrived at that stage they seem bound to furnish a valuable and helpful clue concerning the interaction between mind and matter.

CHAPTER XXI

ETHER AND THE SOUL

With Introductory Remarks on Modern Physics

"Which of us, who beholds the bright surface
Of this ethereous mould whereon we stand,
This continent of spacious heaven, adorn'd
With plant, fruit, flower ambrosial, gems, and gold,
Whose eye so superficially surveys
These things, as not to mind from whence they grow."
                    MILTON "Paradise Lost."

There was a time—not so long ago—when electricity was thought of as something vague, mysterious, and immaterial—a mere affection of matter; it was sometimes called a form of energy, sometimes popularly spoken of more specifically as a kind of vibration; it seemed, in fact, like a mere outcome of the properties of matter, which displayed themselves in this singular fashion. Those who studied it closely knew very well that electricity was not a vibration, and not even a form of energy, but what it was they did not know. Some pioneers, greatly daring, spoke of it as a fluid, and the notion of a fluid got into popular language and acquired a sort of spurious definiteness which was unjustifiable and had to be condemned; for the term suggested that electricity was merely an imponderable form or variety of ordinary matter. Heat also was spoken of as a fluid, and the facile term became useless and misleading.

226

ETHER AND THE SOUL

The progress of discovery soon showed that heat was not a form of matter at all; that it really was a vibration, a mode of motion, a form of energy; and we also learnt that without ordinary matter heat as such could not exist. The idea of temperature became rightly associated with the conception of the rapidity of motion of material particles; and the irregular motion of such particles—molecular motions at random, disorganised, and in every direction—turned out to be precisely what heat really is. A useful analogy is sound. Everyone knows that sound is due to the vibration of bodies as a whole—that such vibration disturbs the atmosphere, and that this disturbance when conveyed to our ears is what we call sound. Without a vibrating body, and without a conveying atmosphere or other material vehicle, sound cannot exist: A ghostly or disembodied sound is meaningless. In a vacuum there is no such thing. A vacuum is permeated with absolute silence. Ring a bell in empty space and there is nothing to be heard.

Heat differs from sound in being not due to any vibration of a body as a whole, but a material body is necessary, for heat is a quiver of its ultimate particles. Given a purely molecular disturbance of quite irregular character—then this irregular movement, if vigorous enough and applied by contact to any portion of the skin, is what stimulates the nerves with the sensation of warmth, and is accordingly known as heat.

But what about light? Is that a material vibration too? or is it a projected substance? Is vision due to a bombardment of particles? or is it caused by the impact of aerial vibrations? The answer now accepted is, Neither. Light is not a stream of particles—though that view has been held by influential thinkers,—nor is it transmitted as any disturbance of the atmosphere, nor as any affection of ordinary matter. It is a vibration certainly, but a vibration of something far more perfect and fundamental than any form of matter known to chemists. It is only hindered by matter; it is a kind of disembodied vibration; its home is vacuum. Across empty space it can travel with perfect ease. The filament of a glow lamp is enclosed in a vacuum—in a vacuum as perfect as can be made; for although the vacuum is not really perfect, the lamp would be all the better if it were— *i.e.* if exhaustion of air could be carried to the extreme limit; yet the filament is perfectly visible. The space between the stars is still more empty of matter—as empty of matter as anything we know; and hence it is that their light reaches us with ease across those terrific distances, even though it take a thousand years on its journey. Not until their light enters the gross and matter-containing region of the earth's atmosphere is the free progress of star-light interfered with. Then indeed it encounters difficulties, and visibility may be impaired. To light, matter is an interference, an obstruction, never an assistance. Light is a kind of discarnate heat. And, when the energy is incorporated in matter, heat is what it becomes.

I use the term "light" here to signify briefly every kind of radiation conveyed by the ether at one and the same definite pace. Sometimes a portion of such radiation is called radiant heat, but it is all one essentially, the varieties only differing as bass notes differ from treble ones. "Radiant heat" is a popular not an accurate term, and it is sufficiently erroneous to be confusing. The thing that travels is not heat, but has the power of generating heat when absorbed by matter. But in this respect every sort of radiation is in the same predicament. The kind of radiation longest known to us is that which affects the eye, and so it is permissible to use the term light occasionally in a generalised sense, covering the whole range from X and γ rays to the waves of wireless telegraphy—at least when we do not wish to discriminate. Etherial radiation, however, is the correct term whenever we wish to be precise.

By studying the effects of matter upon light something may be learnt about the properties of both, especially about the properties of matter. Light passing through the difficulties of matter is trained and analysed, and may be made to yield up its secrets, but in itself and apart from matter it is invisible and intractable. Matter helps us to deal with it, but matter is no help to the light itself. And yet it is through or in matter that it originates: in some sense that is true. Every source of light known to us is a piece of matter. By matter apparently it may be generated, by matter it can be destroyed, but not by matter can it be conveyed; nor is matter in the least essential to its existence—only to its accessibility.

A recent discovery is the fact that light is only emitted by matter in quanta, of which no fractions are possible, so that a whole quantum must be emitted or none at all. That is a very interesting and most important fact, having to do with both the generation and the absorption of radiation, but having nothing to say about radiation in transit; that is, the quantum is not concerned with the ether considered by itself, but only with its interaction with matter. Radiation shows no sign of a quantum while in transit, it then consists of continuous waves; only when we experiment upon it with a material instrument does it exhibit any sign of discontinuity. In transit the behaviour of light is explicable in terms of wave motion: interference, diffraction, and the rest go on just as if quanta did not exist; we have to take them into account only during the time when it is being emitted or absorbed by matter.

This enables us to evade some of the difficulties felt by physicists when they try to reconcile the undulatory theory of light with the facts of atomic processes, such as photoelectricity. The quantum in fact contains two factors, one ν for the frequency of the light, the other h for the dissociated unit of angular momentum appropriate to an electron when forming part of an atom of matter. The whole $h\nu$ is the amount of energy involved in the transaction. This is the expression given by Einstein for the energy of a quantum, but he thought of it travelling through space as a unit, for which there is no evidence. The evidence of the quantum of Max Planck is wholly concerned with emission and absorption by matter. As Professor Fleming has said, in other words, because a trough or canal is filled with bucketsful at one end and the water removed from it by buckets at the other, it does not follow that the water flows along the canal in bucketsful, it may flow quite continuously and only exhibit discontinuity when it encounters the discontinuous structure of matter.

Light is known to be a vibration, a particular kind of vibration which is able to stimulate the nerve filaments embedded in the retina of the eye; but what the thing is which is vibrating, the thing which really does transmit light through empty space—about that we may popularly be said to know nothing: for it is a medium which does not appeal to our senses in any way. If we try to experiment upon the luminiferous medium, we fail. It is absolutely intangible and insensible, and yet it has properties of its own; and one of its properties is the easy transmission, at a perfectly definite pace, of that strange tremor or vibration which we call light.

Do we know anything else about it? Yes, essentially and really we know a great deal, but our knowledge has not yet been properly formulated. Suffice it to say that we utilise the properties of the luminiferous medium not only when we send telegraphic messages or drive electric tram-cars, but also when we wind up our watches, or bend a bow, or fire an explosive—yes, even when we raise a weight or bicycle down hill. The strength of materials depends upon it, and half the energy of all terrestrial activity exists not in sensible and ponderable matter, but in this subtle and elusive medium—the medium which transmits light, and which is known as the ether of space.

Light is an affection of the ether. Light is to ether as sound is to matter. What about electricity and magnetism, then? Are they related to the ether as heat is to matter? Is electricity some peculiar affection of the ether of space? So it had been thought. In various vague forms this idea must have been in the back of our minds.

Electricity seemed something immaterial and elusive, only becoming apparent when associated with ordinary matter. In the form of an electric charge on the surface of bodies, in the form of an electric current running through metals, electricity became known and was studied for a century. But whether there was anything substantial about an electric charge, whether there was anything really running along a wire conveying a current, that was greatly doubted and thought improbable. Ideas about it were vague and unsubstantial, notwithstanding a vast accretion of both theoretical and practical knowledge. It seemed a thing almost hopelessly inaccessible, only to be dealt with in association with matter, only tractable in its innermost nature by the power of high mathematical analysis.

But a change has come over the spirit of electrical science; and, largely owing to the genius of men still

living, electricity has put on body, and form, and size,
and mass; it has become corporeal. It is a substance—
a fluid if we like to call it so,—it consists of particles,
not material indeed, but corpuscular. It has become
concrete and substantial, though it remains inaccessible
to our senses. Electricity is no form of ordinary matter,
but suggests itself as the raw material out of which
ordinary matter is composed. It is still refined and
subtle, but it is no longer elusive. The corpuscles can
by indirect and most ingenious means be counted and
weighed and measured. They crowd on the surface of
a charged body, and constitute its charge. They rush
through a wire conveying a current, and constitute the
current. They whirl in almost infinitesimal orbits, and
constitute what we know as magnetism. They swirl and
change in speed or in direction of movement, and thus
excite in the ether the specific disturbance which appeals
to our eyes as light. That is how light is generated, by
the changing movements of electric corpuscles. And
when light is absorbed or mopped up again, it is to the
astonishing evolutions of those same corpuscles that its
energy is entrusted. The period of indefiniteness and
recondite puzzledom has come to an end, and the period
of definite conceptions in this department of physics
has begun.

Subject to all the laws of time and space, fully amenable to the laws of energy, largely the source of terrestrial
energy, governing all the manifestations of physical
force, at the root of elasticity and tenacity and every
other static property of matter, the ether is just beginning to take its rightful place in the scheme of physics;
and the etherial corpuscles—the specks of modified
ether which we know as electric charges—are beginning
to be recognised as substantial little entities, in terms of
which are to be interpreted the very constitution of
gross and ponderable matter. They are the units of

which atoms are composed. They, and not primarily the atoms of matter, appear to be the ultimate foundation-stones of which the material cosmos is built. Electricity is no longer vague and mysterious, it has become unexpectedly accessible and tractable; many things about its constitution remain to be discovered, but these little substantial nuclei, with their definite and ascertainable movements, are dim and ghostly no longer.

Not that these electrical units or electrons displace or supersede the material atoms. They help to explain them, but they co-exist with them. Electrons are themselves material, and by their organisation have acquired some elementary properties of matter. And yet they can be considered separately, and they are not ordinary matter. We know both matter and electricity, there is room for both—one involves the other, though they can be detached and considered separately; and if ultimately our conceptions attain a more comprehensive unity, it will be the electric and not the material unit which ultimately and most fundamentally survives. Electric charges, composed of modified ether, are likely to prove to be the cosmic building material. Meanwhile atoms and electrons co-exist, and for practical purposes are separable; even if only as a swarm is a unit distinct from its constituent bees, or a wall something different from its constituent bricks. The differences may be greater than what is here suggested: we do not at present clearly know. The two kinds of electricity seem to combine to make one kind of matter: or rather to make many kinds—to combine in different ways so as to make all the chemical elements.

But, besides these peculiar portions—the modified minute electric units—there is the great bulk of undifferentiated ether, the entity which fills all space and in which everything material occurs. A duality runs through the scheme of physics—matter and ether. Matter appeals to our senses, but the unmodified ether makes no such appeal; it is so inaccessible that its existence even has been denied. No one can deny the existence of matter, at least not on common-sense grounds; common sense can easily deny the existence of the ether. Yet one is as certain as the other, and in all the activities of the cosmos the interaction of the two entities is essentially and clearly involved. All kinetic energy belongs to what we call matter, whether in the atomic or the corpuscular form; movement or loco-motion is its characteristic. All static energy belongs to the ether, the unmodified and universal ether; its characteristics are strain and stress. Energy is always passing to and fro from one to the other—from ether to matter or *vice versâ*—and in this passage all *work* is done.

Now, the probability is that every sensible object has both a material and an etherial counterpart. One side only are we sensibly aware of, the other we have to infer. But the difficulty of perceiving this other side—the necessity for indirect inference—depends essentially and entirely on the nature of our sense organs, which tell us of matter and do not tell us of ether. Yet one is as real and substantial as the other, and their fundamental joint quality is co-existence and interaction. Not inter-action everywhere and always, for there are plenty of regions without matter—though there is no region without ether; but the potentiality of interaction, and often the conspicuous reality of it, everywhere prevails and constitutes the whole of our purely mundane experience.

Many other instances can be quoted, in science, of the gradual progress from vague and indistinct conceptions to something tangible and concrete. I have drawn an example from electricity, but a biologist could readily draw upon other branches of knowledge. For instance, it is popularly known that *malaria* simply means "bad air," and the disease was attributed vaguely to the unwholesome atmosphere of marshes. But the source of the disease has been traced to the bite of a creature which breeds in water, a familiar insect which can be seen and heard and crushed. Indeed, the whole treatment of disease in general has been revolutionised by Pasteur's discovery that it was usually the result of poison secreted by actual little organisms visible in a high-power microscope. The vague and indeterminate always tends to become the substantial and definite, as knowledge advances.

Now, applying all this, as a sort of parable, to the probable progress of psychical research, I must foresee a little. I must look ahead. It is a *tendency* which I recognise, not an accomplished fact.

Our ideas about the nature of the soul have hitherto been for the most part vague and tentative; the term "soul" has been used more or less apologetically, and the corresponding reality has been difficult to grasp. Some reality was felt to be underlying the term, but it was too vague and indefinite to be susceptible of satisfactory definition; it seemed too subtle and elusive for everyday use. The term was usually avoided. Body we knew, and spirit we might in sort imagine—for clearly the matter of living things is controlled and arranged by something. But soul, what was that? Well, I foresee a time when the term soul will be intelligible, and I think it will be found that soul is related to the ether as body is related to matter. I suggest that it will turn out to be a sort of etherial body, as opposed or supplemental to our obvious material body. That is what I foresee as lying in the path of the progress of discovery. We shall find, I think, that we possess, all the time, a body co-existent with this one that we know—a body essentially substantial and related to space and time, not really transcendental, but yet in no way appealing to our present senses. Intangible and insensible, it may yet exist; and if it exists it may be detachable and capable of separate existence. It will be the etherial aspect or counterpart of our present bodies, but more permanent than they. For there is no property in the ether which suggests ageing, or wear and tear. These, and other temporal disabilities, such as fatigue, imperfect elasticity, friction, dissolution, belong always to an assemblage of material atoms. No imperfection of any kind has yet been detected, or even suspected, in the ether of space. In so far as it is a fluid at all it is a perfect fluid: its elasticity and all its properties are perfect.

It may be asked, How can we acquire any knowledge of a supersensuous thing? So long as any portion of the ether, or an etherial body, interacts with matter we can know something of it—not of it itself, but of what it can do. While still in the flesh we shall probably only know our etherial counterpart through its interactions with matter. Directly these cease, it passes beyond our ken; but it exists just as really as before. Indeed, freed from the disabilities and imperfections of matter it can lead a less distracted and livelier existence; it enters on a less troublous career, its vehicle is no longer this gross matter, with its heaviness and its dissipation of energy, but the free and perfect ether; of which, in some way yet to be discovered, it appears to be composed.

Hypothetically and tentatively I am inclined to postulate a duplicity of construction even in inanimate objects, for without it there could be no unity or coherence or any individual object at all—nothing but a dust of disconnected atoms. Ether is the medium of cohesion, it is that which holds the particles together, and it is that which is strained when the particles of an elastic body are displaced. The material particles are only shifted in position: the joining substance—the ether—is strained, and makes them spring back. The ether and the matter are interwoven.

Naturally it is only among animate objects that any psychic significance attaches to either the material or the interwoven etherial aspect; and the amount of psychic significance to be attached to any particular specimen of animated matter depends on its grade in the scale of existence. But whatever spiritual or mental affinities can be found in one, presumably belong equally to the other—that is my hypothesis; and co-existence of the two aspects is involved in, and renders possible, the familiar terrestrial life which we are aware of here and now.

As to the psychic relation: I assume, as a working hypothesis, that whenever a psychic element acts at all, it normally interacts with both the material and the etherial aspects; it may be suggested that perhaps it makes use of one for the activities we call conscious, and utilises the other for the activities which at present are unconscious. I do not press that, nor think it very likely. But I do postulate interaction with both matter and ether, and not only with one, as hitherto contemplated by perhaps the majority of philosophers. Whether the etherial portion can actually continue its psychic connexion, apart from any material counterpart, is a question for evidence, not for dogmatism. But assuming the possibility—assuming that a withdrawal of the mental or spiritual guiding agency from the material aspect, so that that ceases to be animate, need not mean withdrawing control from the etherial aspect also—a good many things become vaguely explicable, or at any rate can be brought into a state ready for contemplation; though it is clear that to produce any manifestation to our present senses some elements of the material fraction must be temporarily restored.

The bringing in of the ether into the scheme of psychics, as it has already been partially brought into the scheme of physics, is the work which I feel sure is lying ahead for generations of men. Then—when a serious beginning in this direction has been made—the term "soul" will acquire a definite and clear connotation; no longer will the idea of a spiritual body seem vague and indefinite and difficult of apprehension—there is nothing indefinite about future existence—soul will no longer be regarded as a term to be avoided, but will become as real and recognisable, as concrete and tractable, as are the corpuscles of electricity. The interactions which are possible between the matter of this planet and the etherial bodies or souls associated with spiritual intelligence will then be understood; and with this knowledge, under proper regulation, a new power will be gained; and this new power will be utilised and put into action. The meaning of human life, and the puzzles which surround it, will then become clearer and more intelligible. The obscure communications and strange movements which are now studied or experienced in spiritualistic circles, and which by some are thought to be miraculous or impossible—so impossible that the reports of them usually excite ridicule—will gradually take their place in the orderly scheme of recognised science. Such things as travelling clairvoyance and reciprocal dream experiences may become intelligible. Gratitude will surely then be felt to those early pioneers who in past centuries maintained their form of truth in spite of persecution, and testified to what they had known in face of undeserved contempt. Ultimately the subject will emerge from its dark and difficult period

ETHER AND THE SOUL

—a period clouded with traces of superstition and obstructed by well-meant but antiquated prejudice— and familiar intercourse across the veil or gulf of death will become sufficiently common to prove an untold blessing to the human race.

PART FOUR

THE EVIDENCE FOR SURVIVAL AND ITS MECHANISM

CHAPTER XXII

EVIDENCE FOR AND MECHANISM OF SURVIVAL

"The splendours of the firmament of time
    May be eclipsed, but are extinguished not;
Like stars to their appointed height they climb,
And death is a low mist which cannot blot
    The brightness it may veil. When lofty thought
Lifts a young heart above its mortal lair,
    And love and life contend in it for what
Shall be its earthly doom, the dead live there,
And move like winds of light on dark and stormy air."
                    SHELLEY—"Adonais."

Thus we find in the ether a reasonable home or habitat for those who have lived in association with matter for the allotted time, but whose existence is by no means terminated by the sloughing off of the material body. The ether seems well adapted to serve as the new vehicle for their use in a continued existence. But this semi-physical theory of their conditions would not have occurred to me if I had not been acquainted with proofs that they actually did survive and continue; so it becomes necessary to give some idea of what that evidence is. It is very bulky, and will be found in innumerable books of testimony by members of bereaved families: the few samples I shall quote are only to be taken as typical of the kind of evidence to which we attach importance. More and probably stronger cases will be known to many readers.

PRELUDE

In discussing the evidence for human survival it is natural to select and recount a number of cases in which information is given by dead people about some incident known only to themselves and subsequently verified. There are a fair number of episodes of this kind embedded in the *Proceedings of the Society for Psychical Research*, and indeed in many books too; and some of them to a student of the subject are certainly striking and of considerable interest. But, in order to be effective, the communications must deal with very trivial incidents personally connected with the deceased person, and not likely to be known to anyone else. Historical events, or occurrences of national importance, are much more likely to be known to living people, and would thus be rendered nearly useless as evidence of individual survival. But a trivial episode, say the narration of some action secretly performed by the deceased person while still in the flesh, and not known by any of his surviving relatives or indeed by any other person, would be much more suitable for a crucial posthumous test. It cannot be easy to think of information that is not or may not have been known by some surviving mind; while anything like public events that have appeared in the newspapers are obviously possibly known to a lot of people, and are thereby ruled out of account, and have to be ignored by any stringent critic, because the hypothesis of telepathy from the living person who was aware of the facts must be pressed to the uttermost before the critic can pass the evidence as entirely crucial and satisfactory.

This circumstance, the discovery that telepathy or mind-reading can occur between one living individual and another—though itself discovered in the course of our researches, and not yet universally admitted by all students of science—has become a curious obstacle or bugbear to those who wish to establish individual survival. Communications purporting to come from deceased people are, as must now be known, exceedingly numerous; but for the most part these communications, though often specially characteristic of the person supposed to send them, and often very satisfying to relatives and intimate friends, are not in themselves entirely demonstrative to the scientific critic and to outsiders. A great mass of evidence of this kind has now accumulated, and much of it is of a kind that is superficially conclusive. Affectionate and family messages giving names and other details about the family are very common, and are naturally taken to represent what would be expected of the personality concerned. They show distinct signs of surviving personality, memory, and affection; but mainly they all refer to matters well known to the recipient of the communication, and accordingly are open to the accusation of mind-reading, i.e., to the hypothesis that the information contained, though entirely unknown to the medium employed as the means of transmitting the messages, can be supposed extracted from the mind of the sitter by some process unconsciously exercised by the medium. So that, though at first bowled over by the naturalness of the conversation held through a medium with the deceased person, the relative or sitter, who is aware of the discovery of telepathy, is subsequently afflicted with doubts as to whether the messages really came from the person ostensibly represented as sending them.

I think myself that this hypothesis of a widespread power of telepathy is very much exaggerated, and that most of the communications indicative of the surviving memory and natural affection of the deceased communicator, when they come through a good medium, are really what they purport to be. Certainly if the same kind of messages were received through a telephone from, say, a lost but living relative at a distance, the clue to his identity would be recognized as complete, and no question about his genuineness would be entertained. When the tones of the voice are reproduced, when the memories are characteristic of the person, and when the incidents recalled are sufficiently trivial to be known to only one or two people, then the proof of identity is for all ordinary purposes accepted, and no further doubt need harass the recipient of such a telephone message. But when, as frequently happens, a message with the same strongly marked characteristics is received from one who is known to be dead, the fact to be established is so extraordinary that scepticism is bound to be pressed to extremes, and every normal possibility exaggerated, before the evidence can be considered sufficient to establish the fact beyond all doubt.

Occasionally, however, as will be seen after a study of the records of the S.P.R., it happens that a deceased communicator is so anxious to establish his identity that he has arranged before death to do some trivial thing in private—to hide, for instance, some object in a specified place—and then that he takes pains afterwards to give information about his action to some survivor, in order that it may be verified. I will give one instance of this kind of occurrence merely as a sample. It is an experiment which few people take the trouble to make, but which might be made by more if their attention was directed to the reasonableness of achieving such a proof.

CASE OF THE HALF-BRICK

A small but significant episode is reported, after first-hand enquiries, by Dr. Richard Hodgson, well known as an emissary of the English S.P.R. in America, in the Proceedings of the S.P.R., Vol. VIII, pp. 249–251. It had been narrated in the *Religio-Philosophical Journal* for 31st January, 1891, but Dr. Hodgson made enquiries and

EVIDENCE FOR SURVIVAL

got into correspondence with the sister of the deceased communicator, a Mrs. Finney, who wrote to him as follows:—

"ROCKLAND, MASS.,
19 *April*, 1891.

"For weeks and months before my brother [died] we conversed freely on the subject of spirit communion and such matters, and one morning he requested me to bring him a small piece of brick, also pen and ink; he then made two marks on one side and one on the other with the ink, then, breaking the brick in two, gave me one piece, telling me at the time to take care of it, and some way he would hide the other piece away where no one but himself would know, and after leaving the form, if possible, would return in some way and tell me where it was. I could then compare them together, and it would be a test that he could return and communicate, and *my mind* could not have any influence over it, as I did not know where he put it.

"After he left the form our anxiety was *very great* to hear and learn all we could of communicating with spirits, and for months we got nothing satisfactory.

"We then commenced sitting at the table at home (mother and myself), which we did for some little time; at last it commenced tipping, and by calling the alphabet spelled out where we could find the piece of brick that he put away—that was the way we got the test."

In a later letter she adds, under date 3rd May, 1891:—

[The message we received was] " 'You will find that piece of brick in the cabinet under the tomahawk.' I went to that room and took the key, unlocked the cabinet, which had not been touched by anyone after he locked it and put away the key. There I found that piece of brick just as it had spelled out, and it corresponded with the piece I had retained, fitting on exactly where he broke it off the piece I had. It was wrapped in a bit of paper and tucked into a shell, and placed in the bottom of the cabinet *exactly under* the tomahawk, as well spelled out by the alphabet."

And then, in response to further questions, she adds on May 11th:—

"The piece of brick was entirely concealed in the shell, so that it could not be seen from outside of cabinet. It was wrapped in a piece of paper stuck together with mucilage and tucked into the end of the shell, then a piece of paper gummed over that, so that nothing was visible from the shell. The shell was on the lower shelf of the cabinet, and only the top of the shell was visible outside the cabinet."

And later again:—

"I send you by express a box containing the letter and shell with the piece of brick. I have placed one piece in the shell just as it was when I found it, so you can see how nicely it was concealed in the shell. The papers that were around it then are worn out. You can retain them if you like, as I do not care for them now. To me it is a positive truth that he did communicate to us, and our minds could have nothing to do with it. . . .

"The shell was placed on the same shelf with the tomahawk, and no other shells on that shelf. It was placed with the open side down, and the tomahawk stood directly over it. I cannot say why he did not tell us to look inside of the shell. We started to look as soon as he told us. It was in the cabinet under the tomahawk. We did not wait for any more to be said."

To this Dr. Hodgson adds:—

"The shell is a large Triton, about ten inches long. The piece of brick was wrapped in folds of soft paper and tucked deeply into the recess. Another piece of paper was then gummed around the sides of the shell in the interior, so as absolutely to prevent the piece of brick from falling out. When I received the shell from Mrs. Finney and looked into the interior and shook the shell violently, there was nothing to indicate that the shell contained anything but the piece of gummed paper.

"The piece of brick in the shell weighs about one and a half ounces, and the piece of brick retained by Mrs. Finney weighs about two and a quarter ounces. The shell with the piece of brick and paper wrapping weighs about eleven and a half ounces.

"Mrs. Finney also forwarded me the letter written by her brother. The shell and the pieces of brick and the letter are now all in my possession."

EVIDENCE FOR SURVIVAL

The "letter" last referred to by Dr. Hodgson is a sentence which the deceased had deposited with his sister, sealed, in order later to be posthumously deciphered; the sentence itself is so trite and commonplace as not to be much evidence, though to the sister it was likewise conclusive, since he gave through the table the very words written in the sealed envelope.

The posthumous letter incident is narrated thus:—

"One more little incident I will mention, for to me it is as valuable as the other. He wrote me a letter (about the time he gave me the piece of brick) and sealed it, saying at the time it was not to be answered, but the contents of the letter to be told. I got that in the same way as I did the other, by calling the alphabet and the table tipping. It was these words:

"'Julia! do right and be happy.—Benja.'

"That was correct. Just the contents of my letter. I have no particular objection as to giving my name, for I have stated nothing but the truth. (Signed) JULIA A. FINNEY."

Another instance of a somewhat similar kind is described by Mr. E. F. Benson under the guise of fiction, for which, however, I have reason to believe that there is some substantial foundation, near the end of his book called *Up and Down*, where the deceased communicator, a young man killed in the War, had hidden in what he called a 'cache' three distinct objects, and then after death told his friend what they were, so that he could verify the statement.

CASE OF THE OLD SCRAP-BOOK

An authentic case is narrated in full by Mrs. Sidgwick in the "Proceedings" of the S.P.R., Vol. XXXI, pages 253–257, but this statement I must greatly abbreviate. Its date is 1917, and the information was obtained through the mediumship of Mrs. Leonard and her control "Feda" by a lady (Mrs. Hugh Talbot) who was a complete novice at the subject, and who treated it at first as an absurdity not worth following up. The communicator was her deceased husband, who was anxious to convince her of his identity; so he tried to tell her of an old note- or scrap-book of his, about which she knew nothing, but which after a hunt she succeeded on finding at the back of a dusty shelf. The statement of the communicator was that she would know the book when she found it by finding in it a diagram of Aryan and Semitic languages, with lines branching out and connecting the different entries. A number of details were given with much reiteration and signs of anxiety lest his wife should not understand and not take the trouble to verify. The communicator also asked her to read page 12 or 13 of this book, as it contained something that interested him now very much. She did not at first take the steps necessary to find the book; but ultimately she was urged by a niece to do so, and reports as follows:

"I wanted to wait till the next day, saying I knew it was all nonsense. However, in the end I went to the bookshelf, and after some time, right at the back of the top shelf I found one or two old notebooks belonging to my husband, which I had never felt I cared to open. One, a shabby black leather, corresponded in size to the description given, and I absent-mindedly opened it, wondering in my mind whether the one I was looking for had been destroyed or only sent away. To my utter astonishment, my eyes fell on the words, ' Table of Semitic or Syro-Arabian Languages,' and, pulling out the leaf, which was a long folded piece of paper pasted in, I saw on the other side 'General Table of the Aryan and Indo-European Languages.' It was the diagram of which Feda had spoken. I was so taken aback, I forgot for some minutes to look for the extract. When I did I found it on page thirteen. I have copied it out exactly."

On page 13 there is an extract representing the feelings of a person at the point of death, or rather the sensations of a man passing through death, and seeing the faces of

EVIDENCE FOR SURVIVAL

people whom he knew had already gone over. It is an extract from an anonymous book called *Post Mortem*, published by Blackwood in 1881.

The lady adds:—

"I do not attempt to reproduce the diagram of languages, which is complicated, but Feda's description of it as having lines going out from a centre is correct; this branching out from points and from lines happens repeatedly."

She says:—

"I cannot account now for my stupidity in not attaching more importance to what Feda was trying to say about the book . . . [but] it was only my second sitting. I knew nothing of mediums and the descriptions seemed so endless and tedious."

There are other cases of the same general character, which can be looked up in the 'Proceedings' of the S.P.R.:—(also the careful and accurate record referred to on page 275.)

Vol. VIII, page 200 and page 238;

Vol. XVII, pages 181–182;

Vol. XXXV, page 511;

Vol. XXXVI, pages 303–5; and then the strange and well evidenced finding of a will on page 517, *et seq.*, which I have narrated in my book called *Why I believe in Personal Immortality* (Cassell).

This last is a sample of the kind of episode when the object to be found is regarded by the deceased as of some value, and therefore stays in the memory, and can be used afterwards as a test when the opportunity occurs for giving posthumous information about it.

Such episodes have been known for generations. Some of them, occurring in the life of Sewdenborg, were recorded by the philosopher Kant, when comparatively a young man, though subsequently he seems to have taken a dislike to the subject, and did not adhere to his first enthusiasm, so that he spoiled his little book, *Traüme der Geisterseher*, by a kind of hedging appropriate to the incredulity of those times.

But now I apprehend that I ought not to cite too many narratives which, whatever their significance may be, have a superficial air of triviality. I just want to insist that the triviality is part of their merit, and to say that, if I myself find an opportunity of communicating I shall try to establish my identity by detailing a perfectly preposterous and absurdly childish peculiarity or habit, which I have already taken the trouble to record with some care in a sealed document deposited in the custody of the English S.P.R. with supplementary information deposited with the L.S.A. I hope to remember the details recorded in this document, and relate them in unmistakable fashion. The value of the communication will not consist in the substance of what is communicated, but in the fact that I have never mentioned it to a living soul, and no one has any idea what it contains. People of sense will not take its absurd triviality as anything but helpful in contributing to the proof of the survival of personal identity.

There are a number of evidences for survival accumulated by the S.P.R., much studied by the leaders of that body, and more favourably regarded by them than these sporadic instances of communications about family affairs. They are of a different character, and consist of evidences of scholarship, given through comparatively uneducated mediums, and the whole ingenious system of cross-correspondence which originated from some of our deceased members on the other side who were acquainted with the difficulty raised by telepathy and took special steps to overcome it. These communications are remarkably interesting and cogent when studied, but to make these clear as evidence would require a considerable time to be given to each of them, so I have limited myself to the simplest kind of evidential communications.

What I want now to do is to tackle the question of the theory of survival from a theoretical point of view, as an outcome of my lifelong study of physics. The statement will involve some repetition of what I have said before, but its importance will justify me.

When we consider the question of Survival from the physical point of view we are up against the ancient problem of the connexion between mind and body. The body is certainly made of matter, but matter is inert, it never does anything, it is completely controlled by the forces acting upon it, which forces exist in the empty space surrounding the atoms. Energy only makes itself manifest by its effect on material bodies, but its main existence is in space. We have no sense organ for perceiving energy itself, our senses tell us nothing but matter. We can see the results of energy as expended upon matter, but we have no direct apprehension of the energy. We are not acquainted with anything in the universe save by its effect upon matter, and that is the origin of our tendency to philosophic materialism; we are liable to doubt whether things not apparent to the senses can have a real existence, though there is no justification for such a doubt.

We cannot understand the activity of the material universe without taking energy into account, and this energy exists in the space between the particles. Matter is discontinuous, consisting of isolated particles; they are connected only through space.

Animated matter differs in no respect from every other kind of matter, except that it is subject to animation. So when we say that life only exists in a material organism we ought to say that life only manifests itself in association with such an organism, and that when it is

dissociated from matter we know nothing of its existence. We have no right to say that it is extinct. All
that we know is that it is no longer manifest, it has gone
out of our ken. But the same may be said of every
form of energy: in itself it has no power of becoming
known to us but by its effect on material bodies. A
body under the action of life can do many things, can
initiate spontaneous movements, can build up an
organism, can operate on the physical universe, and leave
structures behind it of interest and beauty, but it is not
the material body that does these things; they are due to
the life or animation of the body.

If then we can adduce any evidence that life or mental
activity exists in space, and only sporadically makes itself evident by some material activity, the state of our
present knowledge of physics renders your acceptance of
the fact entirely harmonious. We have to do no violence to our physical conceptions if we admit the fact of
survival. Life and mind never were functions of the
material body, they only displayed themselves by means
of the material organism. The organism was not essential to their existence, but only to their display, that is to
our apprehension of them. If they ever find means of
operating in a novel or unusual manner on a physical
organism, then they may still manifest their continued
existence: and that is exactly what they do. Why should
we decline to receive the evidence?

Telepathy shows that mind can act on mind without
the use of any bodily organs, hence certain people may
have a faculty of apprehending a spiritual world direct;
and this may account for genius and inspiration. This
has been well argued by F. W. H. Myers, and I shall not
labour it now.

If you have evidence of the existence of a spiritual
world, a world of help and guidance and sympathy, then
you can hold to it in face of every denial of the

MECHANISM OF SURVIVAL

materialists, who can only base their denial on the absence of any sensory stimulus to their material organism. Such a world may exist all round us, and yet can only be spiritually discerned. The faculty of discernment does exist in some people, and their positive evidence overweighs a wilderness of negation from people whose perceptions are limited to the bodily senses. One of the most elementary forms of discernment is (rather absurdly) called psychometry. An object put into their hands may convey more information than the senses can give. A psychometrist can tell something of its history, something of its association, something of its possessor. By special faculty they can tell far more than could be arrived at by chemical tests. They can tell, for instance, that a bit of stone has formed part of a pyramid, or that a ring has taken part in a scene of slaughter, or that a piece of writing or a drawing has been done by a certain person normally quite unknown to them, and can even tell what the circumstances of that person were at the time, and what they were doing. The faculty of psychometry is apparently unknown to or disbelieved in by my friend the Bishop of Birmingham. If he took it into consideration he might modify—not his main doctrine, but some of his expressions concerning the beliefs of others.

The existence of a spiritual world throughout the depths of space is becoming to me a great and fundamental, even a physical, reality. The manifestation of that world in connexion with material organisms on one or other of the planets is a comparatively trifling and temporary episode, of great importance doubtless in the history of evolutionary development, but our real existence is not dependent on a material organism. Our spiritual and real home is in the ether of space.

Chemists and biochemists are liable to limit themselves unduly to the purely material aspects of things. A chemist's business is to deal with matter in its various forms: that is his job, and he need not be expected to go beyond it. A physicist takes into account the ether as well, though he may for a time prefer to call it space. He is not limited to material particles, but studies the fields of force which connect them and make them active. The psychologist goes further still, and studies the action of mind. I would I could say that the biologist is a student of life, but at present the tendency is for him only to study animated organisms and their behaviour, limiting his attention to what is manifested by the material processes brought about by life, and not thinking that life has any existence apart from its instrument of manifestation. We shall never understand the universe by attending to matter alone and ignoring everything which makes it active and interesting. We cannot even understand the bending of a steel spring nor the fall of a raised weight without implicitly taking the ether into account. We are continually making experiments on the ether and realizing the consequences of its abundant qualities. If we make the assumption that it is a physical vehicle for life and mind too we are only extending our generalization in the same direction.

A supplementary and semi-physical treatment of Survival is now becoming possible; a treatment which is well calculated to replace the old materialistic view that man had only a material body, and that when that body died and decayed, the animation, the personality, and the individual, necessarily ceased to exist. It is also well calculated to replace the popular idealistic notion that any spirit which survives the death of the material body must survive in an entirely disembodied condition and be out of relationship with the physical universe. Many people suppose that it then belongs to another order of existence, or, as some would say, of non-existence; that it is likely to be free from any relationship even with

Space and Time, and must have departed entirely out of our ken; so that communication or intercourse is no longer possible, until perhaps at some future day when the material body shall have been somehow resuscitated and restored to its old function, in glorified form, so that the spirit can resume its active control. That this superstitious idea has been prevalent is testified to by popular modes of expression, such as:

> "On the resurrection morning all their dead the graves
> restore:
> Father, mother, sister, brother meet once more."

This depressing notion of future existence—if it can be called existence in the interim—is not a scientific or psychological view at all; but it has been the religious view through mediaeval times, hymns and liturgies are saturated with it, and it continues to this day the chief representation of what by strictly orthodox people is meant by Survival.

A modern theory which seeks to provide the emancipated spirit with any kind of organism related to the physical world, might thus be ranked as a return to a modified form of materialism. For though, when properly understood, the view I advocate ought to emancipate us from materialistic bugbears, and although it wholly condemns the idea that flesh and blood or any particles of terrestrial matter are revivified and inherit Eternal Life, yet popular ignorance of what is meant by the Ether, and of the certain fact that the Ether is a part of the physical universe and has definite properties which can be experimented on and ascertained, may well suggest all manner of difficulties in understanding the hypothesis I am trying to expound. Wherefore it will probably be considered unsatisfactory, both by the scientific materialist and by the theologian; possibly also by some spiritualists.

MY PHILOSOPHY.—PART IV

The necessity for some kind of organ or instrument or habitation for an emancipated spirit has been intuitively felt by many inspired writers. The most ancient classical idea was that of a condition rather melancholy— homeless, wistful, shadowy and sad—but this notion was improved upon even in later classical times. "Not unclothed but clothed upon," "God giveth it a body," are modes of expression very familiar to readers of St. Paul's epistles.

The existence of a spiritual body is an idea, in one form or another, at least as old as St. Paul. It has been upheld by some of the Greek Fathers of the Church; it has been vaguely in the mind of many modern investigators; sundry obscure and supernormal facts seem to lend it strong support. And recently an etheric version of such a body has been approved—and, if not inculcated, at any rate regarded as a step in the right direction —by some of the more thoughtful and philosophically minded communicators "on the other side."

What they know by experience is that, though discarnate, they are certainly not disembodied: they feel no more disembodied than we do. They tell us that they still have substantial instruments of manifestation which serve for intercourse among each other, and that it is through this permanent instrument that they are able, occasionally and under certain conditions, to operate indirectly, through our organisms, on the matter of this planet. They operate with more difficulty than in the old days, partly because they have to make use of other people's mechanism; but still, subject to many restrictions, they exert influence in a somewhat similar way, and thereby are able occasionally to know what we are doing; and they claim sometimes to succeed in helping and stimulating us, not only mentally but physically.

Now, although the departed may not understand fully and completely of what their present body is composed, or how they operate on it so as to produce the results they design and aim at, they are still only in the same predicament as they were when here, and as we are now. For we do not know how we control our bodies of matter, nor what the nature of the connexion between mind and matter is. We know that we have muscles and nerves and brain-centres. We can dissect and describe this part of the mechanism. But how a physiological instrument—how any kind of mechanism —can think and feel and plan and will and remember and hope and love, we certainly cannot explain. And probably we never shall be able to explain how such a thing can happen; for the thing to be explained does not happen, it is only imagined to happen through a misapprehension. The truth is that it is we ourselves who really do all the psychical things: we employ our bodies only as instruments for recording and transmitting our thoughts and for exercising muscular action on matter. The body itself neither thinks nor wills nor sees nor feels. It is an instrument, a channel, a medium.

Although full explanations about our method of controlling a body are not yet forthcoming—either on this side or on that—yet those "on the other side" are quite willing to accept the suggestion that their bodies, which to them feel so substantial, and all the surroundings in which they exist, are related to the thing which we here call the Ether, very much in the same way as they used to be related to the familiar thing known as Matter.

Until instructed, we can hardly help thinking of matter as dense, and of ether as tenuous, but that is a poetic illusion associated with the term "ethereal." It is an illusion based on the testimony of our senses, which, as so often happens, have to be corrected by deeper insight into the real nature of things. Matter appeals to us so strongly, not because it is anything but a gossamer-like or milky-way existence in the vast continuity of ether, but because our obvious bodies are made of matter, and because our animal sense organs are specially adapted to existence in association with matter, and give us information about nothing else. Even light, which we *know* is an ether vibration, tells us nothing about itself without study; what it tells us familiarly is—not about light, but—about the material objects which have emitted or scattered or differentially absorbed it. We get this information by life-long, indeed age-long, inherited and instinctive experience. We interpret the luminous indications without difficulty, and we forget the strangely complex nature of the processes which underlie all our channels of information; we only find their true nature out when phenomena are fundamentally analysed and seriously cross-questioned. When we have pursued this line of investigation for many years, we find that the important thing in the physical universe is ether, and that matter is trivial comparison. Yet we can freely admit that matter takes such splendid and beautiful forms that it is worthy of the continued study of generations of scientific men; and we need not wonder that they become so enthusiastic over its properties that they are able to imagine it the sole reality in existence.

SUMMARY

The revolution in physical science which has been going on all through this century has had the effect of directing our main attention away from material bodies and concentrating it on the multifarious happenings in space. In my belief this process will go further, for that is where our real existence lies, and there is our spiritual home.

It is in the interaction of ether and matter that the problems of psychology must find their solution. Our existence is essentially independent of this material organism that we have constructed and use for a time. Matter only serves as an index or pointer, demonstrating the unseen activity all around; and chief among the unseen activities are the etheric agencies guided and controlled by Life and Mind and Spirit.

CHAPTER XXIII

ON THE DIFFICULTY OF PROVING
INDIVIDUAL SURVIVAL

" . . . . ever on the watch
Willing to work and to be wrought upon,
They need not extraordinary calls
To rouse them; in a world of life they live,
By sensible impressions not enthralled,
But by their quickening impulse made more prompt
To hold fit converse with the spiritual world."
WORDSWORTH—" The Prelude."

We are bound to ask whether the proof of individual
survival is complete or whether enquirers may be legitimately diverted by the many alternative hypotheses
that could be said to account for certain kinds of experience such as are often taken as proof of identity.
Proof after all is a question of probabilities. I do not
suppose that a sledge-hammer proof which knocks
down all opposing theories and finally exterminates
them is ever attainable by us even in physics. It has
occasionally happened that a theory has been adopted
and has held the field for nearly a century, and yet
has been modified at the end and incorporated with
some opposition theory which had seemed extinct.
What we need in science is a working hypothesis that
we can test, getting results from it that we can verify,
until ultimately its probability becomes so great that
we may have confidence in it as an approach to
certainty.

INDIVIDUAL SURVIVAL

If we grant the possession by a medium of an extensive faculty for telepathy and clairvoyance, no record of the past is safe from the application of those powers. Yet there is a limit to our credulity in those directions, and, after time and much experience, one may feel that our only escape from accepting the straightforward appearance is to strain the alternative hypothesis unduly. An extensive faculty of clairvoyance can hardly be treated as an extension of the normal faculties of the medium, without assuming the intervention of some other intelligence of whose activity many phenomena contain more than an indication. Evidence available for establishing the existence and activity of *some* intelligence, other than that of incarnate humanity, may be said to amount to proof: but this does not establish the activity of any specified individual.

What we have established, I consider, is the existence of a spiritual world. To establish personal identity in connexion with intelligence is a more difficult problem. The faculty of reading a closed book can hardly be attributed to the powers of a person encased in the flesh and using his bodily sense-organs, it involves something more. And so does the perception of a bodily apparition or representation of some deceased person. The Ancients used to attribute an apparition or phantasm, not to the person immediately represented, but to a messenger from the gods, who could put on the appearance of that person for the purpose of conveying a message. Witness the legend of Ceyx and Alcyone in the Eleventh Book of Ovid's "Metamorphoses," where Morpheus impersonates the dead husband, and gives a message to his sorrowing wife.

It becomes therefore a question which can legitimately be asked whether any plan can be devised whereby personal identity can be proven, or at least rendered more probable as an explanation than any other hypo-

263

thesis that can be suggested. Anticipation of the future has been suggested, but that is more than we ought to expect from any deceased relative of our own. The power may sometimes be exhibited, but it suggests a knowledge more likely to be attained by some intelligence who has had time and opportunity to go beyond common experience, and to transcend any powers possessed by ordinary humanity.

The main crux is the establishment of the personal identity of some quite ordinary person. Facts in his life, or documents recorded by him in the past, would seem to be comparatively useless, since they can be arrived at or deciphered by other means. I want to consider therefore whether there are any steps that could be taken by anyone interested in the subject, and anxious to do his best to prove his own survival, that would be accepted by a severe critic, as crucial or conclusive, or at least as raising the probability in favour of that view to so high a degree that for all practical purposes it amounts to certainty.

Let me therefore take an imaginary case, and see what flaws can be found in it. The test most approved seems to be some personal habit or idiosyncrasy, not too well known or capable of imitation, and preferably not known at all, but recorded by the individual before his death in such a way that the record can be deciphered afterwards and be accepted as reasonable. Suppose for instance that a person whom I will call A has had an obsession since childhood of some absurd childish verse or short poem, which has frequently recurred to him throughout life, but which, being entirely frivolous, has never been spoken about or mentioned to a soul, the insignificance of it being such that A is rather ashamed of it than otherwise, and sees no meaning in mentioning it. He may perceive however that this obscurity makes it valuable as a posthumous record, and he may take the trouble to deposit it in safe custody.

The supposed poem or verse must be quite meaningless, not forming any part of daily life, and not having the least interest even for himself, but yet one which has been so frequently recalled as to be permanently ingrained in his memory, so that he may hope to remember it sufficiently to recall it in full detail even after he has lost his bodily organs and passed through the transition called death; assuming, as we are bound to, that such a state of things is anyway feasible, for if we start with the initial impossibility, no proof can be given us. Suppose further that the deceased A has sufficient sense to perceive that the mere recital of the poem as recorded in his posthumous document would not suffice; for it would have to be assumed that a medium could read that, even in a closed envelope. Let us suppose therefore that he takes pains to enter into further detail, having had a lifelong opportunity of doing so. He has noticed, let us say, that the verse contains seventeen words of one syllable, nine words of two syllables, four words of three syllables, and one of four; and that as a preliminary he gives through several mediums hereafter the meaningless jumble of figures 17, 9, 4, 1, and then in due time amplifies the meaning of these figures, and says what they are intended to signify; and that, foreseeing this attempt that he will make hereafter, he has recorded in his deposited document the fact that he will send these figures, and will afterwards expand their meaning. He might take some further steps, and record why or under what circumstances the trivial verse took such a hold on him. It must not be a verse of any importance, or one that could be called a quotation. It might be a portion of a Bab-ballad, for instance; though even these are in the knowledge of several people, and are too well known to be suitable. I would rather suppose that the thing he has thus thoroughly memorised is quite uninteresting and meaningless, even to himself, that he cannot explain why it has stuck in his memory and been analysed in almost a physiological manner, the details of which analysis could hardly be imagined by the most expert clairvoyant, but which nevertheless, in the case I am imagining, is to be given through several mediums by the person who thus intends to establish his identity, and will likewise be found recorded in the posthumous document that is to be ultimately opened and read.

If this experiment could be carried out thoroughly, both by the deceased person who thinks the object aimed at sufficient to justify him in taking all that trouble, and by the survivors who have been entrusted with the document and who must be supposed to act wisely,—not to be in a hurry, but to wait until they get the assent of A himself as to what they think they will find in the deposited document, before they open it,—then to me the hypothesis that that individual is still functioning, and that the messages really come from him and from no one else, seems to me the only one that meets the case. There may be loopholes for scepticism even in such an imaginary case as that. It requires rather an exceptional concatenation of circumstances to make the test possible. But if that would not be a proof that would satisfy a reasonable critic, then I don't see how anyone anxious to prove his identity hereafter could proceed with the task. At any rate I mean to try something of the kind in my own case when the opportunity occurs, and I appeal beforehand for reasonable treatment.

CHAPTER XXIV

ON THE REASONS FOR THE NON-RECOGNITION OF PSYCHICAL RESEARCH BY THE MAJORITY OF THE SCIENTIFIC WORLD

"So the blind doctors in the country of the blind told those who saw the sunset that they were the victims of a reversionary hallucination."

J. ARTHUR THOMSON.

In psychical investigations we must insist on the necessity for care and caution in making and recording observations. We must be on our guard not to be deceived. Fraudulent phenomcna are the devil. No building can be erected on rotten foundations: fact must be supreme. But apparently facts alone are not adequate to secure scientific recognition. Where possible, observation must be supplemented by experiment; and the whole must be united by theory. This last is a most important aim. Until we have a theory, or at least a working hypothesis, facts are apt to be neglected as troublesome and discordant, and are liable to be denied.

The science of Astronomy is for the most part built on observation rather than on experiment, though a good deal of the spectroscopic part has to be interpreted in terms of previous or subsequent experimentation. But the prestige of Astronomy does not rest on observation of the canals of Mars, nor even of the moons of Jupiter or the phases of Venus. It rests upon the magnificent theory which has been evolved about the simpler kind of phenomena, such as the motions of the planets, which brings them all within a comprehensive law. Briefly the prestige of astronomy is due to the completeness of quantitative theory, and the agreement between calculation and observation. Psychical research is obnoxious to science not because the results are obtained by observation, but because at present we do not know the laws of the phenomena, and have no generally agreed upon theory on which to explain them.

Before the theory of cyclones, (which, though now perhaps a matter of controversy, was certainly useful in its day, and is still useful), meteorology was more nearly in the position of psychical research. The weather was not experimented upon, it was observed; any number of local and temporary facts were accumulated, but as long as they were not strung together into a theory, they excited but little scientific interest; and if it could be called a science at all, meteorology was a science with a low prestige. The phenomena of thunderstorms were accepted because they were common knowledge, or rather, common experience; but until the electrical nature of the disturbances was ascertained by Franklin, they aroused next to no scientific interest. And even now, the phenomenon of ball or globe lightning, which has been often testified to, is only accepted by a few meteorologists, who have been forced to their conclusion by comparatively ignorant and mainly unqualified sporadic testimony; its occurrence attracts but little attention because, so far, ball lightning cannot be explained.

Some scientists urge that true "science" does not begin until metrical considerations enter and quantitative measurements can be made. This is an exaggeration, but there is a tendency in that direction. Until the conditions of a phenomenon, such as the contagion of a disease or the hatching out of an egg, can be explored to some reasonable extent, the phenomenon attracts very little scientific attention. Unexplained facts divorced from theory are only attended to by a few inquisitive or open-minded pioneers.

The fact that the majority of scientific men hold aloof from any class of phenomena is no evidence against their reality, but is a natural consequence of the state of our ignorance, or the smallness of our knowledge about them. When so many things require attention, and yield a promising field for investigation, selection has to be made, and the less promising are left out in the cold.

The distinction sometimes drawn between controllable or controlled, and uncontrollable, phenomena seems to assume that when an experiment is thoroughly controlled by some qualified observer, his report is to be implicitly trusted; whereas if the evidence depends on testimony of others, it is all suspect. This distinction, this selection of a single trustworthy person as conclusive, is not illustrated or borne out by history; for even when an observer has had all the threads in his hand, his statements have been discounted; no one is or ever has been regarded as finally trustworthy. To take an example:—

Sir William Crookes made a very simple experiment with the famous medium D. D. Home, in which he had arranged everything to his own satisfaction, without any interference from the medium. He had a mahogany board with one end on a table, the other end supported by a registering spring-balance. He got Home, sitting at the table, to put his fingers on the fulcrum or table end of the board, sometimes with an intervening vessel of water in which the fingers dipped. When the time was ripe, the far end of the board went down, the scale registering a fair amount of force, the board being depressed as if by an inexplicable increase of weight. After repeating this a good many times, he reported the fact to the Royal Society, under the title "Psychic Force," and invited authority to see it. But the eminent Secretary of that Society, the great mathematician Sir George Gabriel Stokes, after some correspondence, maintained that the result was mechanically impossible, and declined to be a witness of it. In other words, the testimony of a good and famous experimenter about a simple though incredible result, entirely controlled by himself, was not accepted. It seemed contrary to the laws of Mechanics; and I see no reason for supposing that it would be accepted now.

On the other hand, the distribution of responsibility among several co-operative observers, though it may increase complexity, should not of itself invalidate either an experiment or an observation. A multiple observer is not trusted; true, but then a single observer is not trusted either. Hence it is possible to lay too much stress upon the plausible idea that an observation, to be valid, must be entirely controlled by one person. That indeed limits the area of responsibility, but it does not guarantee scientific acceptance.

I take it that the real strength of our position lies in the phenomena themselves, namely that they can be experienced by a number of people, and by one person after another, as time goes on, until the cumulative evidence becomes overpowering. Moreover I hold that what are called "uncontrolled" experiments are not to be despised. For sometimes, under what may be called, and are really, less than strict conditions, phenomena occur of such vigour, and of so simple and striking a character, that they overcome suspicion and constitute their own demonstration. I take as instance another experience narrated and verbally testified to by Crookes; to the effect that on a well-lighted dining-table, in full view of the assembled company, two glass objects on the table rose in the air, and tapped against each other, so as to tap out a sort of message in accordance with the usual code. Whether there was a message or not, the movement of untouched objects in full light, observed by a number of people, might be such as to overpower the demand for strict control, and render disbelief among those present impossible.

Unfortunately such circumstances are rare; and even when they do occur, outsiders can only be informed by testimony; and they prefer to turn that testimony down, rather than accept what they consider an impossibility.

My view is that no record of any experiment can be made watertight and free from suspicion, if lurking grounds for suspicion exist in a critic's mind. I believe that any experimental communication to the Royal Society, even of the most orthodox scientific character, could be permeated with a sort of humorous doubt, and holes picked in the record, by anyone who thought it worth while; "Ah ha! What was the assistant doing just then?" and so on; on the same sort of principle as the "Historic Doubts Concerning Napoleon Bonaparte," which I believe was perpetrated and published by Whewell nearly a century ago, in illustration and so to speak reprobation of doubts about the historicity of a far more important Personage,—doubts then in vogue among German theologians.

It may be truly said that an incident of the kind narrated by Crookes, however apparently conclusive, has evidently not been accepted, otherwise there would be no doubt about physical phenomena. I agree: I do not even know that a serious record was made of that particular semi-domestic occurrence: I only heard of it orally from Crookes, I am not vouching for it. But my point is that it would not have been any more accepted if every member of the company had been searched and handcuffed and chained up or controlled mechanically in any desired manner, with all their utterances recorded on a gramophone, and all their movements on revolving drums. Elaborate precautions are desirable, but they are no real safeguard, nor any guarantee of good observation. Too much faith may be put in mechanical control; indeed the more complicated it is, the more does it occupy the attention of the observer, which ought to be concentrated in other directions. If the mechanism is controlled by assistants, doubts about confederacy inevitably arise.

Indeed if I were a conjurer or fraudulent performer, anxious to deceive a man of science, I should like him to have as much apparatus to attend to as possible, for then there would be no need for me to take trouble to distract his attention. Sufficient distraction would be automatically supplied. Perhaps it is hardly realised by non-experimenters how much attention apparatus needs. It is quite liable to go out of order at a critical moment, and while it is being put right, or tinkered with, there must be opportunities for preparation for subsequent display. Electrical tests, and recording mechanism, sound impressive to a layman, but there are always difficulties about erecting apparatus in a strange place among unusual and unsuitable surroundings. A regular permanent laboratory, where experiments of all kinds are habitually carried on in an orthodox and customary manner, would suffer less from this disability; but a genuinely sensitive physical medium, induced to sit amid these strange surroundings, could hardly help being affected by them, with some not improbable inhibitions. Hence, apparatus, except of a very simple kind, may favour fraud instead of preventing it, and at the same time may have a deleterious effect on genuine phenomena. I am not against apparatus, of course, but urge considerations of common sense in its use.

The contention, now frequently and plausibly made, that no observation is worth anything except under the most stringent conditions, is neither practicable nor wise in all cases. Conditions effective in one direction may be defeated by deficiency in another; we cannot always tell beforehand what precise phenomenon is going to be produced. Precautions taken against "telekinesis" are ineffective against an "apport," and *vice versa*. Moreover it is always possible for an outsider, reading the record in which something incredible has seemed to happen, to assume that some precaution was after all neglected, and that if he had been there, things would have been different.

Nothing is likely to carry real conviction except the cumulative effect of first-hand experience, of various kinds, under a great variety of circumstances. I would rather myself concentrate on ascertaining the laws of a given kind of phenomenon; feeling sure that among all the repetitions necessary to that end, any fraudulent procedure must sooner or later leap to the light. It does not follow that in any specific case fraud has actually been used, merely because some loophole for its occurrence is afterwards thought of. There should be evidence that it actually occurred, and not merely a suspicion of its possible occurrence, before anyone is denounced. Indeed there are occasions when the actual achievement of a result by normal means, in a possibly unconscious state, must be treated as part of the phenomenon to be observed, without moral reprehension or emotion of any kind. Each incident of that kind can then be guarded against in the future.

MY PHILOSOPHY.—PART IV

Neither complete control nor complete observation is easy—perhaps not even possible; hence it may be said that in general no single experiment can be treated as conclusive. The fact that, in spite of abundant testimony, the world is not yet convinced, is sufficient proof of that. There are occasions when a comparatively casual observation may be at any rate of preliminary value. To this class belong the spontaneous cases, say, of apparitions, or premonitions, or of unexpected occurrences generally. From the nature of the case they cannot be prepared for; no prearranged precautions can be taken. And yet such spontaneous occurrences are frequently testified to, and the testimony is admitted as coming within the scope of, and securing publication by, even the S.P.R.

That Society has done good work in Telepathy of various kinds. Ingenuity has been used, and a multitude of precautions have been taken, some of them by those on this side, some apparently by those no longer on the membership roll of the living; yet, even in the case of the highly estimated cross-correspondences, the physical possibility of collusion can hardly be excluded. Whether for that reason, or for some other, they have clearly had no appreciable effect on the scientific world, nor so far as I know on any other responsible group. We must admit that the recorder or student of any unusual phenomenon, even though a man of standing,—a Cabinet Minister or a fellow of the Royal Society,—will not be treated by the rest of the scientific world as immaculate, and beyond suspicion of either consciously or unconsciously exaggerating or decorating his record,

What then is the remedy? In default of an acceptable theory, which would put the subject on a different footing, I see none other than patience and perseverance, coupled with scrupulous care; not forcing the pace but

ORTHODOX DIFFICULTIES

biding the time, and leaving it to the facts themselves to attract competent attention and gradually overpower hostility, by sheer weight of evidence. I doubt not that in due time the facts and their revolutionary meaning will become part of accepted knowledge. But first they must run the gauntlet of what is after all pardonable and even complimentary scepticism; for the scepticism is due to the novelty of the facts. No, not exactly to their novelty, for in a sense they are ancient enough,—but to their vast significance, and to the upheaval of ideas which must follow a general acceptance of their truth.

One of the chief bugbears or difficulties legitimately felt about the reality of communications, purporting to be from the other side, is the possibility of telepathy between the sitter and the medium, so that it has to be assumed that everything known to the sitter can be obtained from the medium when entranced. To meet this difficulty many efforts have been made; but one that must be specially referred to is a series of sittings conducted by Miss Nea Walker for the benefit of a bereaved lady who had lost her husband. To show that it was he who was sending messages, Miss Walker went to a medium before being told much about this lady, before meeting her, and before she knew anything about the nature of the relation between her husband and herself. The result of these sittings is recorded in a book called "The Bridge," (Cassell), by Nea Walker, in which the details of the sittings are given together with subsequent annotations by the widow and other friends, the whole being carefully recorded and being well worthy of study. The details seem long and complex but it is pre-eminently a book for students. It is a challenge to sceptics.

CHAPTER XXV

ON THE APPARENT ELEMENT OF CAPRICE INTRODUCED BY THE SPIRITISTIC HYPOTHESIS

An anonymous writer in the "Proceedings" of the Society for Psychical Research expresses the difficulty of the spiritistic hypothesis thus:—

". . . Regarded as a scientific working hypothesis, spiritism does not seem to me to be a very hopeful avenue of investigation. The spirit hypothesis has a delusive appearance of simplicity, but so also had Kepler's hypothesis of guiding angels. And how remote this was from the complex reality of Einstein's description of gravitation! In fact, if these supernormal mental phenomena depend on the whims and caprices of departed spirits, then I for one despair of ever being able to discover any law and order in them."

Undoubtedly there is some difficulty, in our present state of comparative ignorance, about specifying or formulating the spiritistic hypothesis in any precise and so to speak scientific manner; for it is an appeal to the activity of unknown agents, acting by unknown methods, under conditions of which we have no experience, and by means of which we are unaware. We get into touch, or appear to get into touch, with these agencies only when they have affected material objects, for instance someone's brain, thereby stimulating muscles so as to produce results which appeal to our normal senses.

276

APPARENT CAPRICE

But the admission that we cannot understand how agents work does not justify our denial of the existence of such working. A good deal of modern mathematical physics is in the same predicament. I now quote from another book of mine called *Phantom Walls*, to the end of this chapter, something that ought to be said in this place, instead of merely making a reference to it.

The fact that we sometimes have to postulate an unknown agency does not justify our attributing anything capricious to that agency. We are ignorant of how the gravitational agent acts, but we know that it acts in accordance with law and order, so that the results can be duly predicted. Einstein's view (if we may call it Einstein's, though in one form or another it must have been vaguely held by many) is after all *not* so very different from Kepler's asserted hypothesis. What Kepler meant by "guiding angels controlling the planets" (assuming that he used that phrase) I do not know; but I am sure he meant nothing capricious. He must have meant that an unknown something guided the planets in their path; and that is a paraphrase of the modern view. The something is now often spoken of as a warp in space—acting as a sort of groove. In so far as Kepler postulated something in immediate touch with a planet and acting directly on it, he had what now appears to be truth on his side; his thesis being perhaps nearer the ultimate truth, though far less practically useful, than Newton's delightfully simple quantitative expression for the indirect action of a distant body.

In order to illustrate direct guidance by contact action, we may cite the familiar example of a gramophone needle, which automatically reproduces a prearranged tune, simply by following the path of least resistance. What else, after all, can an inert thing do? That is the meaning of inertia. Animated things are not inert: they need not take the easiest path. A man may

277

climb the Matterhorn for fun. But inanimate unstimulated matter never behaves with any initiative or spontaneity: it is strictly inert. Atoms never err nor make mistakes, they are absolutely law-abiding. If they make an apparent error, if a locomotive engine leaves its track, we call it a catastrophe. All machinery works on that principle; every portion takes the easiest path.

It is true that to get a coherent result there must have been planning and prearrangement. Certainly! In all cases of automatic working, whether biological or other, that must be an inevitable preliminary. But explorers of the mechanism will detect no signs of mental action by their instruments or their senses. To infer a determining or controlling cause they must philosophise. Indeed, we may go a step further and emerge from the past into the present:—A wireless set talks like a gramophone, and to one accustomed only to gramophones it would seem barbarously superstitious to urge that in the wireless case some (possibly whimsical and capricious) operator was actually in control. Statements may be unpalatable, and yet be true.

Now return to gravitation. Planets behave *as if* they were attracted by the sun. That is certainly true. But what is attraction? A train is not attracted to its destination; lightning is not attracted to a chimney; but it gets there none the less, by continually taking the easiest path. So it is with a planet. Indeed, one might say that everything inert takes the only path open to it, it has no option. The law is a sort of truism. But the principle, once recognised, has been formulated into a clue; the Principle of Least Action can be expressed mathematically. Once postulate that, and the behaviour of the inanimate portions of the cosmos can be accurately deduced.

The modern statement that the planets move along the line of least resistance, or the easiest path, makes their motion rather closely analogous to that of a railway train guided by the rails. The path and destination of a train are determined by the continual direct influence of the rails, which make it easier for the train to travel in the right direction than to jump them and go astray. We might, if we chose, admit that the path was laid down or determined by the mentality of the surveyors and designers of the route; but a Martian spectator with partial information might still wonder at the apparent intelligence which guided one part of a train to Manchester, and other part to Liverpool, in accordance with the wishes of the passengers or the labels on the coaches. If told that an invisible guardian angel switched over the points to produce this result, he might resent the suggestion as absurdly unscientific and preposterous; as on a purely mechanistic view it would be.

After having studied trains for some time, our spectator might begin to notice the novelty of a motorcar. His first tendency would be to look for the rails in that case also; and, finding none, he might superstitiously but correctly surmise that a guardian spirit was guiding the car to its destination. In this case, moreover, further experience would soon persuade him that he had to allow for an element of caprice. But even that is not fatal to the truth: he need not throw up his hands in despair. As soon as we introduce the activity of life and mind we get away from mere mechanism, and the results are not easily formulated or predicted. The activities of an animal cannot be expressed in mathematical terms, and yet animal instincts and behaviour are subject-matter for scientific investigation. It is assumed that they obey laws of some kind. Science is not limited to the accurate data and laws of mathematical physics: and to claim that a hypothesis is unscientific because we cannot formulate it completely, or because we do not understand the method of working, or even because there is a certain amount of capriciousness about it, is more than we have any right to claim. Anthropology and sociology are less advanced sciences than physics and chemistry: they have to get on as best they can, with a profusion of data, and with the inevitable complications appropriate to live things. Let us not be put out of our stride by the fear of retaining, in modified form, some of the animistic guesses of primitive man. Experience may lead us, as it led him, to contemplate stranger modes of existence, and more whimsical phenomena, than our long study of mechanism has led us to expect. We must put aside prejudice, be guided by the evidence, and strive for truth. The superficial simplicity of materialism has served us well, as a comprehensive covering, for many centuries, and we have made good progress under its protection; but it is beginning to get threadbare and inadequate, it is not coextensive with reality, and unsuspected influences are peeping through.

To sum up. A working hypothesis can be followed up and developed rationally without being metrically exact in its early stages. The important question about the spiritistic hypothesis is not whether it is simple or complicated, easy or puzzling, attractive or repellent, but whether it is true. Its truth can only be sustained or demolished by the continued careful critical and cautious method of enquiry initiated by the S.P.R. under the Presidency of a guiding spirit or guardian angel called Henry Sidgwick, with the active (and I believe continuing) co-operation of Edmund Gurney and Frederic Myers.

CHAPTER XXVI

THE WHOLE ORGANICALLY CONSIDERED

And now, having cleared the ground so far, let us consider what is this spiritistic hypothesis about which there is so much trouble. The trouble is caused partly by our philosophic views. If we are unwilling to admit that we are spirits here and now, utilising material bodies which we have automatically constructed for the purpose, then probably any form of spirit hypothesis is unwelcome and perhaps meaningless. If we are nothing but material mechanism—if a collection of organic molecules, merely by reason of their chemical complexity, can develop certain functions and reflex actions characteristic of what we call life—if such mechanism can become aware of itself, can admire, and plan, and acquire a sense of controlling its own actions, if in fact matter constitutes our essential existence— then the hypothesis of an animating spirit may well be considered unscientific and grotesque, and one which ought to be abolished from the scientific vocabulary.

But many philosophers now urge, and I think reasonably urge, that materialistic philosophy has broken down, and does not cover the whole ground. Materialistic mechanism is true as far as it goes but it is not the whole truth. The attempt to make it the whole truth is a natural outcome of the astonishing success of mathematical physics during the last three centuries. This powerful science at first dealt completely with moving particles exerting mechanical forces upon one another according to any prescribed law, then extended itself to rigid bodies on the one hand, and perfect fluids on the other, and so gradually introduced elasticity, viscosity, the theory of the conduction of heat, and the molecular movements associated with gaseous and other states, including all the vibrations responsible for sound. This science, penetrating to the actual forces at work and analysing every detail of their acton by a marvellous method of mathematical deduction, seemed to form a complete and satisfactory and ideal scheme. It began by reducing astronomy to an admirable system of law and order; indeed Newton and Laplace seemed to be initiating the last word on the detailed elaboration of the solar system, and events could be predicted centuries ahead. Astronomy seemed approaching a kind of perfection. The complacency of nineteenth-century physics was remarkable. Naturally an attempt was made to explain all phenomena in the physical universe in terms of molecular interaction: that was the ideal set before himself by Newton, that ideal has been followed up ever since. So far as the inorganic world is concerned we might hardly expect to dive deeper, for though the explanations really leave a good deal of mystery from the philosophic standpoint, it is the kind of mystery to which we have grown accustomed in daily life, and usually ignore.

The most advanced sciences throughout the nineteenth century flourished on this basis, and set an example which other sciences tried to follow. The ambition of physiologists has been to apply these same laws to organic and living structures, and to work out the behaviour of the animal and vegetable kingdoms on a physico-chemical basis, and no other. Certainly there were naïve experiences which constantly suggested difficulties, and seemed to demand something more;

THE WHOLE ORGANICALLY CONSIDERED

but it was hoped that these difficulties were of the kind that could be overcome by further study, and by a still closer understanding of the chemical and physical processes involved. There seemed to be nothing in the universe but matter in various forms of motion; and by a thorough study of matter and motion it was hoped that the whole of nature might be understood and explained.

To quote approximately from Professor Whitehead's book, *Science and the Modern World*, Chapter IV.:—

"The eighteenth century was the age of reason; healthy, manly, upstanding reason; but, of one-eyed reason, deficient in its vision of depth. . . . Voltaire was typical of the virtues of his century; he hated injustice, cruelty, repression . . . and he hated hocus-pocus. . . . But if men cannot live on bread alone, still less can they do so on disinfectants. So the age had its limitations; yet some of its main positions are still defended in the schools of science. . . . The seventeenth century had provided a perfect instrument for research. The triumph of materialism was chiefly in the sciences of rational dynamics, physics, and chemistry. . . . Nothing fundamental and new was introduced in the eighteenth century, but there was an immense detailed development. Special case after special case was unravelled. It was as though the very Heavens were being opened, on a set plan. . . . In this century the notion of the mechanical explanation of all the processes of nature finally hardened into a dogma of science. The notion won through on its merits by reason of an almost miraculous series of triumphs in mathematical physics. Newton's *Principia* was published in 1687, Lagrange's *Méchanique Analytique* in 1787, and Clerk Maxwell's *Electricity and Magnetism* in 1873. Practically a century between each. Each of these three books introduces new horizons of thought affecting everything which comes after them."

Fifty years later, however, there has begun a revolt; a revolt, strange to say, led by the mathematical physicists themselves. It had been perceived that a study of matter alone was inadequate, and that the behaviour of even the simplest molecules could not really be understood without attention to the properties of the space around them. In fact it was found that empty space was endowed with physical properties. Those properties cannot be investigated directly: we still have to attend exclusively to matter in all experiments and observations, but the material behaviour is found to be secondary and subordinate to the behaviour of empty space. Newton seems to have strongly suspected something of the kind, but at that time he could not make much headway in this apparently more speculative direction; and his theory of matter particles acting on each other by unexplained forces at a distance was so satisfactory from the mathematical point of view, and gave such exactly verifiable results, that any further treatment in the philosophical direction seemed unnecessary, and at any rate had to be postponed.

The revolt against the concentration of attention on matter alone was effectively begun by Faraday during the first half of the nineteenth century, and has been going on at intervals ever since. Throughout the whole of his masterful treatment, and so to speak creation, of the science of electricity, Faraday insisted on the subordinate part played by material bodies, especially in the phenomena of electrostatics and the electric field. He pointed out that charged conductors were merely the terminals or boundary of an electric field, that inside them the field was non-existent, that it ceased at the boundary, and extended unbroken throughout apparently empty regions between the visible and tangible bodies on which alone experiments could be made. Their behaviour was a sign or token or consequence of the unknown reality which was going on in the space between them. The intervening space might indeed be full of insulating material, but might equally well be entirely devoid of matter. Space itself had what he called a specific inductive capacity or dielectric coefficient, of a nature and value unknown: unknown to this day. Space is modified by the presence of matter, but its properties are not material properties. The primary happenings occur in vacuo.

Much the same idea had already been promulgated in the case of light or radiation. Light is not transmitted by matter, but by space. Matter plays quite a subordinate part. Light travels with the greatest ease and simplicity through the emptiness of space, and is only retarded or interfered with by the most transparent matter. All matter contains some trace of opacity; and by opaque matter light is merely destroyed, and its energy turned into heat; but space is perfectly transparent. The radiation which reaches us from the sun and stars has traversed millions of miles of empty space without the slightest loss of energy. We only *detect* it by its effect on matter when it has arrived: it then affects the retina of our eyes, our photographic plates, and the surface of our skin, setting up the chemical and other changes with which we are all familiar. In particular it operates on the green parts of plants, and thereby renders possible the whole of vegetation. All that we see in a wooded landscape is due to energy which has arrived through empty space, and represents a storage of that energy in visible and tangible form. The energy has as it were become incarnate in matter. A plant or a tree is an incarnation of solar energy; and the complete understanding of vegetative processes is impossible without taking this space energy into account. Just as a charged body was to Faraday the termination of an electric field, so a vegetable organism is the termination of a luminiferous field. The ether may properly be called luminiferous or light-bearing, for it conveys radiation without displaying it. Space itself is dark: there is nothing in it that can affect the senses. It may be full of what we sometimes call ether tremor; but nothing is exhibited, nothing is perceptible, unless a particle of matter is introduced. It is only matter which appeals to our senses, and it is only material objects that we see.

Thus it is through matter that we become aware of the universe; but we need not allow ourselves to fall into the blunder of therefore considering that matter is its most essential feature, or of confusing a phenomenon itself with its index and result. That would be rather like imagining that the deflection of the needle of a galvanometer was the essential thing about an electric current. It would be confusing the manifestation or sign of a thing with the thing itself. The movement of a piece of iron may demonstrate the existence of a magnetic field, and the field can be explored and investigated by the kind of material motions which it causes, but moving iron is nothing like a magnetic field: it is only the sign or index of it. We shall never understand the nature of the field by attending to its demonstration alone; and we should merely stultify ourselves if we supposed that we understand magnetism or electricity by merely studying the movements and rearrangements of matter. Yet that is what we are tempted to do when we are studying the behaviour of living organisms. All we can observe is the motion of matter; and we are liable to imagine that some of those movements constitute life. They are the sign or manifestation of life: they are not life itself. Movement of particles in a brain are very different from "thought."

Essentially the same order of ideas holds even in gravitational astronomy. The movement of the planets demonstrates what is going on in empty space; those movements are the index or sign of the real phenomenon. A gravitational field exists between the worlds, and we should know nothing of it except for their motions. We can attribute those motions to a mysterious force which they exert one upon another; but Newton more than suspected, and Einstein has elaborated, the idea that that force is merely a sign or index—some might say a simulacrum—of some unknown condition existing in space, and that it is to variations in that space condition that their regulated movements are really due. Each particle moves from instant to instant along the line of least resistance or easiest path open to it. It is the space in immediate contact with the particle that guides it; and though we may truly say that it moves *as if* it were attracted by a distant body, we know that that is only a mode of expression—there is much virtue in an "if,"—and that it is the state of the gravitational field in touch with each particle which controls its motion and thereby demonstrates its own existence. Some of what Faraday claimed for an electric field can be extended to a gravitational field also, though there are many important differences.

To step from these long known and comparatively simple examples to the spontaneous movements of a living organism, say an animal, is a very big step, and not one to be undertaken lightly; nor can we deal with it with anything like the same satisfaction or fulness of knowledge. For the science of biology is comparatively in its infancy, notwithstanding its immense range and the enormous field open to its classification and investigation. I would, however, direct attention to the possibility that in so far as an organism surpasses mere mechanism,—in so far as not all its actions are reflex, in so far as it thinks and contrives and plans, and is guided by anticipations of the future and memory of the past, instead of being immediately obedient to present impulse like a planet or a molecule,—I suggest that the organism is the index or demonstration of something beyond itself, something which, though it may for a time be incarnate in matter, has its more real and permanent existence in some other region. Whatever this animating essence may be, it makes no direct appeal to our senses, and is only displayed or demonstrated by its effect on organised material.

In all the cases that I have dealt with it is matter that we observe. The underlying cause or motive power is beyond our immediate apprehension, and is a matter of inference. We infer the properties of an electric field from its effect on what we call charged bodies. We infer and investigate a magnetic field from the behaviour of iron and other substances. We try to arrive at what the electric current really is from its various influences on matter. And Einstein is leading us to infer a curious warp or modification of space from the effect it has on the perceptible sensuous portions of the universe. So also I would hope that we might gradually infer and investigate the nature of an animating spirit from the behaviour of the organism on which we presume it acts.

DEFICIENCY OF A PURELY MATERIALISTIC VIEW

Matter has no initiative. Every particle moves as it is impelled, without plan, aim or intention, just a thoughtless drift—though the result may be satisfactory or beautiful when appreciated by an intelligent spectator. It is instructive to stand on a bridge and contemplate a brawling stream flowing below; for one can realise that every particle is accurately obedient to external forces, and that the whole pattern is consistent with the equations which mathematicians have laid down and worked out. The mental satisfaction in such complete comprehension of an intricate mechanical pattern is very thorough, and has an æsthetic value of its own. To a purely scientific mind this grasp of a problem has replaced and rendered unnecessary the more superficial enjoyment of light and shade and colour and gleaming brightness and glowing depths, which an artist takes delight in and transfers to canvas. This difference, this not inhuman but ultra-human contemplation of the scientific man, strikes deep into the prevalent view of the universe. As says Whitehead, in *Science and the Modern World*, page 24:—

"The particular conception of cosmology with which the European intellect has clothed itself in the last three centuries . . . presupposes the ultimate fact of an irreducible brute matter, or material, spread throughout space in a flux of configurations. In itself such a material is senseless, valueless, purposeless. It just does what it does do, following a fixed routine imposed by external relations which do not spring from the nature of its being. It is this assumption that I call 'scientific materialism.'" . . .

Further on he says:—

"The biological sciences are essentially sciences concerning organisms. During the epoch in question, and indeed also at the present moment, the prestige of the more perfect scientific form belongs to the physical sciences. Accordingly, biology apes the manners of physics. It is orthodox to hold, that there is nothing in biology but what is physical mechanism under somewhat complex circumstances" (p. 144).

"One unsolved problem of thought, so far as it derives from this period, is to be formulated thus: Given configurations of matter with locomotion in space as assigned by physical laws, to account for living organisms" (p. 58).

If this turns out impossible, then living organisms can no more be accounted for or explained on purely mechanical principles than can the ether or the properties of space. We may have to explain mechanism in terms of organism—not *vice versa*. Animation seems likely to be a fundamental thing, not reducible to something else. It may be that animation when properly understood will turn out to be of fundamental importance. Already it has begun to invade the physico-chemical field. During the latter half of the nineteenth century many processes thought to be due to molecular changes in inert matter were traced to the agency of life, *i.e.* of minute creatures operating in accordance with biological laws. Everyone now knows how the work of Pasteur revolutionised the theory of fermentation, of putrefaction, and disease generally, by showing that operations which had been thought to be purely chemical or molecular were really biological and organic. The activity of ultra-minute organisms was found responsible for all these phenomena, and the notion of organism as more generally representative or typical of the processes of nature began to attract philosophic attention.

The whole of nature might be likened to an organism of which we study the functions. On this view, the vital thing is not the structure, but the function. An organism guides and controls its own workings; it operates on and uses matter, and in that guidance the secret lies. The working of the whole is analogous to the working of our own bodies controlled by an animating principle which may be called soul or spirit.

CHAPTER XXVII

THE SPIRITISTIC HYPOTHESIS

"There they discoursed upon the fragile bar
That keeps us from our homes ethereal;
And what our duties there."
                                        KEATS, "Endymion."

Very well, then, the spiritistic hypothesis in its simplest
and crudest form is that we are spirits here and now,
operating on material bodies, being, so to speak, incarnate in matter for a time, but that our real existence
does not depend on association with matter, although
the index and demonstration of our activity does. We
demonstrate ourselves to our fellows only by means of
the material organisms that we have unconsciously
constructed and utilised for the purpose; hence if the
organism is damaged our manifestation becomes imperfect, and if the damage is serious we may have to quit
the organism and remain normally dissociated from
matter. Our activities, on this theory, are supposed to
go on as before, but now presumably in space; and only
when we manage to re-establish some temporary
connexion with matter are we able to make any sign, or
supply any demonstration, of our continued activity.
This is the spiritistic hypothesis, called into existence
to account for a large number of otherwise inexplicable
facts of observation and experiment, *i.e.* of concrete
experience.

Without such demonstration and observation the
truth might not have been known; and such demon-

stration might have been impossible. For even granting some such view of our essentially spiritual nature,—as at any rate plausibly analogous to the other examples of dematerialised activity,—the truth might have been, what many religious people seem to think it is, that once we have lost connexion with matter, that loss was irrecoverable. Religious people seem to think that we are transported or transmuted into a totally different kind of existence, far beyond mortal ken, from which we can make no sign to survivors. They assume that we have lost interest in one another, that we are absorbed in higher things, and hope that those left behind will not disturb us or make any attempt at interrupting us in our new rôle of continuous religious exercises and constant adoration!

I say the truth *might* have been that once the connexion with matter was terminated, it was terminated for good and all; that the consequences of terrestrial action alone survived; and that just as the dead were no longer accessible, so neither could they be assisted by our thoughts or our affection, nor could they have any guiding or helpful influence on those left behind. So completely do some good people think of the departed as in a state of passive rest, that they might as well be in the grave. There are, indeed, some who think that any kind of revival and reunion will only be possible when in some extraordinary way those mortal bodies are resuscitated. "On the resurrection morning soul and body meet again." Those physical instruments seem to them so important that the particles are supposed to be stored until they can be again utilised in some future embodied existence, a millennium, perhaps a million years, ahead. This old, almost superseded, but traditional belief in the resuscitation of the discarded mechanism only serves to show how deep-rooted is the tendency to confuse the indicator with the phenomenon

THE SPIRITISTIC HYPOTHESIS

itself, and to imagine that apart from matter existence is impossible. Whether the physical analogies now adduced are of any use to people capable of such beliefs is doubtful; they know too little about space to be impressed; the popular idea of empty space is sheer emptiness and nothing else.

Yet we have learnt that matter is acted on wholly by the influences which reach it from space. Inert matter would never show any sign of activity, nor could it change its state of motion if left to itself. The changes that we observe are wholly due to the action of space upon it. I want to extend this idea, derived from gravitation, cohesion, electricity, magnetism and light, and include the less known and yet familiar activity called animation. That matter can be animated we can most of us admit, though we know not how it is animated, or what the process of animation is: that space may be animated, too, must be regarded as a new idea. But it is not unreasonable; for just consider:—Long ago it would have seemed absurd to say that space had any physical properties, that light was a function of space, that electric, magnetic and gravitational fields were demonstrations through matter of something going on in space, that the very cohesion between the particles of a solid is due to some entity in space. But to physicists these various properties of space are becoming commonplace; and for myself I venture to extend the conception to animation also. I do not venture to define spirit, save as the animating principle on a higher grade. On a lower grade it might be called soul or mind; and on a still lower grade merely life, which to me seems the rudiment of mind. But whether we are able to define it or not, we all know in some rough sense what we mean by the term. It is the basis of Descartes' philosophy "*cogito ergo sum.*" Whatever else he knew or did not know, he knew that he could think. And it is that thinking, idealising, aspiring, hoping, loving part of ourselves which I wish to suggest by the name of spirit.

My doctrine at present is that this transcendental, immaterial entity needs and always will need something physical—physical, not necessarily material—for its manifestation, that it never is really without a "body," even though it be discarnate. The mechanism of flesh which was utilised here was indeed temporary, but that was never its primary mechanism. The primary physical mechanism associated with spirit is not gross matter, spirit can only interact with matter under difficulties for a time; its real permanent existence is in the freedom of space, with an etherial mechanism, whose properties do not appeal to the senses, and therefore are at present beyond our ken. Our only mode of investigation at present is limited to the occasions when spirit may for a time re-establish communication with matter, in defiance of the popular prejudice that such re-establishment is impossible.

I said that truth *might* have lain in that direction,— the direction of impossibility of communication,—but as a matter of fact our experience shows that it does not. For those who have studied obscure phenomena know that under certain conditions utilisation of a borrowed existing organism, or even apparently the reconstruction of a temporary material body, is possible, and that through this singular use of a discarded method of manifestation, demonstration of continued existence has become real. Communication is not entirely cut off, the departed do not soar entirely out of our ken. By special effort and under special conditions connexion with matter can be re-established, and thus effects are produced which do appeal even to our bodily sense organs. One would not have expected that: but we must be guided by the facts. That the facts point in that direction is obvious. That they conclusively prove

THE SPIRITISTIC HYPOTHESIS

that deduction,—in the sense that they can be explained in no other way,—is what remains for us to enquire into and make sure of. If other hypotheses are successful, then by all means let them be tried. But to be successful they must meet *all* the facts; and to my mind every other hypothesis sooner or later breaks down.

Basing my conclusions on experience, I am absolutely convinced not only of survival but of demonstrated survival, demonstrated by occasional interaction with matter in such a way as to produce physical results. These effects may be accomplished through the loan of other organisms, submitted to the temporary control of an alien intelligence; that is the commonest way. Communion may be and apparently is achieved in more directly mental telepathic fashion also. There are doubtless limits to the possibility of interaction with matter after our familiar organism is left behind; but those limits are what we have to ascertain: we cannot lay them down *a priori*. It may be that naïve experience will lead us nearer to the truth than the most recondite speculations of philosophy. The truth may not be so complicated as some would have us think. Complications are probably due to imperfect knowledge. A simple hypothesis may be quite near the truth, even if we cannot formulate it completely. We cannot formulate our own activities completely, yet it is common enough to adduce the activities of an incarnate spirit: for instance, it is considered quite simple to say that some person has brought a message, or that some other person has removed an object, or that yet another has obeyed a request, and done us a service, and given us a helping hand. These occurrences are familiar enough, though how they are accomplished we should find it hard to express in detail. Philosophers puzzle over the simplest actions. The utilisation of vibrations of the air for communicating ideas is not really a simple process;

it ought to be surprising, instead of commonplace. In our daily life we trust to naïve experience, and we are not misled. We often do not understand facts, we just grow accustomed to them. Understanding comes later; and not to all. If the facts indicate communication and continued mental activity, then let us not be afraid of accepting them. If we had waited to deal with electrical phenomena until we understood the nature of an electric field, we should be waiting still. If Newton had declined to consider gravitational astronomy till he understood the nature of a gravitational field, we should still be in the Dark Ages of science. We have not to wait, before planning things, until we understand the interaction of mind and body.

The spiritistic hypothesis, pressed to the full, probably involves far more than we can in our highest flights imagine. It leads us into the region of æsthetics and genius and inspiration and theology. But our ordinary daily life is conducted on lower levels, and for them the simple primitive ideas suffice. Struggling and bereaved humanity seeks to learn something of the fate of its loved ones, seeks to be assured that affection continues, that they are not far removed from us, and that reunion will not be postponed to some absurdly distant date. My hypothesis is that they are all round about us, in what we call the ether of space rather than in matter, that inter-communion is still possible, and that simple souls may derive comfort from their intuitive perceptions and naïve experiences, without being deterred by the difficulties which successful concentration on material mechanism for the last two or three centuries seems to raise across their path. After all, it is now found that that material mechanism itself contains more mystery than had been conjectured, and that the full explanation even of it, if ever such explanation is forthcoming, will lead, and already is beginning to lead,

THE SPIRITISTIC HYPOTHESIS

towards an idealistic view of existence, not at all dissimilar from the animistic or spiritistic view of the real and permanent universe here and now.

In brief, we are immortal spirits in temporary association with matter. Probably it is through this bodily restriction and isolation that we become individualised and acquire a permanent personality, which hereafter is able to adapt itself to new surroundings, in accordance with the well-studied biological adaptability of the rest of animate existence.

CONDITIONS OF FUTURE EXISTENCE

Why do people decline to face continued existence of the same general kind as that which corresponds with our experience now? The world as we see it is largely our own interpretation. To a different grade of being the same things might have a totally different aspect. Our apprehension depends on the way we interpret sense indications; and if our interpretative faculty continues, we shall be likely to interpret other surroundings in much the same way. The interpretation and even the kind of perception of nature depends a great deal on ourselves; and an interpretative faculty is likely to continue. But hitherto science has declined to contemplate an immaterial existence, and only a few are willing to suppose that it may be full of concrete reality in which we can feel at home.

This disinclination to face realities of a concrete kind in another order of existence, may be partly due to deference to the science of the seventeenth, eighteenth and nineteenth centuries, which seldom contemplated concrete and commonplace surroundings, and mainly dealt with abstractions. It was the poets who faced concrete existence. Literature as a whole is full of it, for it concerns itself with natural objects in their full display of colour and vivacity, as well as with the caprices and behaviour and responsible or irresponsible actions of human beings in general. Whereas science busies itself with the abstract, and mainly prides itself on metrical abstractions of a non-human character. Even the classifications and generalisations of science are abstract. But life does not consist of abstractions and generalisations. Life consists of full-bodied reality and concrete instances. It finds room both for the dignified and the undignified, the frivolous as well as the serious; thimble-rigging and sweepstakes on the Derby find a place in it. It is not high and dry at all. Humanity can be whimsical and seek odd experiences, and can disport itself, as in novels, in the most un-scientific manner. As scientific men we can ignore these eccentricities, but we cannot eliminate them from nature. The world would be poorer and more prosaic without them.

Perhaps the next world is not the remotely dignified, continuously religious place we have been inclined to think. All the gloom and blackness associated with it—aye and the blazing brightness—may be inappropriate. Death is solemn undoubtedly; but so is birth. Entry into a new state of things cannot but be an important adventure. The world of matter on which we enter at birth is wonderful enough, but it has its moments of frivolity; and I see no reason to suppose that any existence in which we share shall seem to us entirely different in that respect.

CONSEQUENCES

On the hypothesis that we can ever enter into communication with those in that order of existence, we ought to learn from them something of what it is like. I consider that we have learnt something, not very much, but sufficient to carry on with. I do not expect to be much surprised when I get there. So far as it goes, the testimony is in favour of a still continuing full-bodied existence, not indeed of matter, but of something else, which though it does not appeal to our present senses, is otherwise equally real, equally substantial; freer and less hampered it seems to be, but not revolutionarily different. We appear to remain ourselves; and the conditions around remain of somewhat the same order. That at any rate is the testimony, for what it may be worth. And it seems to me fairly reasonable that it should be so. Whatever it may be that has produced this world, has produced that also; and the mere fact that it is of a kind which does not appeal to our present animal senses is really no argument for its completely different essential character. In so far as we remain ourselves, we might expect other things to remain themselves too, or at least to be capable of a somewhat similar interpretation in the light of whatever faculties we then possess.

Any hypothesis which aims at becoming a theory ought to lead to results and if possible predictions of a verifiable kind which can subsequently be tested. The difficulty about the spiritistic hypothesis is that we can only test its assertions by observations and experiments on matter. Only in so far as discarnate humanity can still deal with matter can it make any direct appeal to us. Now humanity already possesses a certain power over matter; and it is solely by their effect upon matter that we are normally aware of other human beings. Hence any display of odd material happenings may have to be hypothetically attributed to the extended powers of known humanity.

It is true that our faculties are not limited to material or sensory experiences: we have ideas and intuitions and aspirations, and can frame conceptions of an immaterial kind. Science for the most part ignores these vaguer powers of humanity; but poetry and literature do not; these immaterial things form most of its subject-matter. They constitute a large part of our real experience: and in a survey of existence they should by no means be ignored. To narrow ourselves down to the conceptions of metrical science is an undue limitation of our faculties. It would mean a shutting of our mental eyes to perhaps the greater part of our experience. Immaterial conceptions are not abstractions, but are real concrete mental experiences of which we are primarily aware. And it is a continuation of these experiences that we mean by survival.

Assertions are made, however, about certain phenomena which can only with great difficulty be attributed to the enlarged powers of living people. Things are moved beyond muscular reach, without magnetism or other methods of acting at a distance to which we have grown so accustomed that we call them "ordinary." Temporarily materialised organisms or parts of an organism are testified to; produced not slowly as in the customary operations of biology, but quickly, and liable also to rapid dematerialisation. The bringing of an object into a closed room is said to be an occurrence that really happens; as if matter were not only porous in scientific doctrinal reality, but absurdly permeable to ordinary observation. Immunity from fire, and many other physical phenomena, are taken as indications that some intelligences of greater knowledge or experience than our own are operating. In that respect we seem to be like savages beginning to get into occasional touch with a more experienced white race. So various physical phenomena, as well as the materialisation and direct voice and photographic evidence, are appealed to as demonstrations of a power altogether supernormal, and from the purely mechanistic or materialistic point of

THE SPIRITISTIC HYPOTHESIS

view supernatural. These things, if they occur at all, do not occur in the ordinary course of nature, the kind of nature contemplated by the Royal Society. Their investigation has only been carried on by Societies formed for that purpose. Nevertheless they seem to form the naïve experience of certain individuals. So on the principle that no experience should be overlooked, the growing body of testimony in their favour ought not to be ignored.

It may be hard to define the limitations of extended human faculty; the objection that we do not know the extent of our own faculties is a valid objection. Meanwhile undoubtedly the dramatic semblance of these occurrences strongly suggests outside agencies; and it may be that that suggestion is a true one. We would not have anticipated that discarnate intelligencies, even if they can send messages, have the power of performing miracles; that is to say, of producing effects which so far as we are concerned would be miracles; but if they are able to do these things, we should not ignore them as impossible, but take pains to examine and verify them. The first need is verification of the phenomena; consideration of how they are accomplished must follow in due time. At least they would form an expansion of psychology; and the expansion to which they certainly seem to point is that the psychical entity responsible for our daily and commonplace actions is not limited to its association with any particular temporary organism, but is capable of sporadic activity even when that organism is discarded.

On the whole, however, the verification of this hypothesis, or our willingness to regard it as the beginnings of a theory, must depend largely on our philosophic view. Attempts are already being made by mathematical physicists at an enlarged view of existence. A concentration on matter which served well enough the eighteenth and much of the nineteenth centuries, is turning out insufficient. The abstractions of science clearly do not cover the whole of concrete experience. And the revolt has begun by diverting attention from matter to the properties of space, or to the properties of whatever substantial entity fills space.

Now all this philosophy may seem, perhaps, disjointed from our main theme. But some philosophy of the kind is essential to it. There are two questions about mind. First, Does mind act on matter at all? And next, Can mind supply for itself any experiences other than those provided for it by the body? If our philosophy enables us to answer these questions in the affirmative, then we can proceed. We are no longer limited to the action of incarnate mind. Mental experiences other than those given by our senses are permissible, and mental actions on matter need not be limited to those we are accustomed to. For getting accustomed to things does not explain them; our own actions on matter are just as mysterious as any of those of which we have doubtful testimony. Are we spirits here and now, utilising and acting on matter, or are we mere mechanism simulating the functions of mind? If the latter is true, as some scientific men hold or try to hold, then the greater part of real experience has to be ignored. Literature and poetry become nonsensical. But if we appeal against the abstractions of science to concrete reality as perceived by us, then the former alternative is inevitable, and a great step towards survival is already taken.

My doctrine involves the primary reality of mind in association with whatever physical mechanism it may find available. Matter constitutes only one of those mechanisms, and indeed only constitutes it in secondary fashion; and by a study limited to matter alone we shall never get the full reality of existence. I hold that all

THE SPIRITISTIC HYPOTHESIS

our actions on matter here and now are conducted through empty space, or rather through the entity which fills space; and that, if our activity continues, it must be continued in the same sort of way and through the same sort of etheric mechanism that we already unconsciously utilise now.

That in brief terms is the spiritistic hypothesis which I proclaim and work on.

CHAPTER XXVIII

THE BEARING OF THE THEORY UPON RELIGION

"I but open my eyes,—and perfection, no more and no less,
In the kind I imagined, full-fronts me, and God is seen God
In the star, in the stone, in the flesh, in the soul and the clod;
And thus looking within and around me, I ever renew . . .
The submission of man's nothing-perfect to God's all-complete."

ROBERT BROWNING, "Saul."

If the spiritual world is a reality interpenetrating this state of things and governing them on the whole for good, the fact surely ought to be known to our politicians and statesmen, and their policy should be affected by the knowledge. A changed attitude is essential to human welfare in the long run. There should be more harmony between the Ecclesiastics and the Politicians. At present the ecclesiastical method is to admit guidance by the spiritual world and adapt its language thereto, without any strong conviction that its confidence is justified by fact. The politicians pay a lip service to this doctrine, but go their own way without making any effective appeal. There is therefore complete dislocation between the ecclesiastical doctrine and its outcome in practical politics. The result is not satisfactory; the difference between theory and practice is too great. I do not know the remedy, but clearly this divagation cannot be sound.

The universe seems to me a great reservoir of life and mind;—realities which I believe exist in space, and which will survive the birth and death of worlds, and continue long after the material universe has run down, if its fate is to run down. Life and mind do affect the material universe in a way which is not open to prediction or calculation, and therefore is open to change in deference to other minds. Laplace's calculator could predict all the behaviour of the molecules, given their positions and velocities and accelerations at any one moment; but he could never attempt to treat in that way the action of a live thing. That has a spontaneity beyond his equations. The physicist ignores live things, keeps them out of his laboratory, does not attend to them, they are too complicated. A biologist attends to them, but only from the point of view of their material structure. To philosophise truly requires more than that. The region of religion is not concerned with material objects, it is concerned with the higher entities of which we have some dim apprehension in ourselves. We have to trust our instincts and intuitions. We infer these higher attributes in human beings; but spiritual or cosmic existence is not limited to human beings; there are many entities which give no material sign of their existence, and which can yet operate on the physical universe.

How do we know about the effect of Mind operating on the physical world? We see it all around us. Mind is an organising arranging principle, sorting and ordering. When life and mind are absent, so that unorganised forces are dominant, operations go on, but they always tend towards disorganisation and chaos; order is shattered, buildings crumble into ruin, refuse accumulates, organisms decay. There is a tendency to return to chaos. When life and mind operate, a reign of law and order begins, things are built up into organised structures, and are prevented from tumbling down; except when inorganic forces are too strong, as during a tornado or an earthquake. Under the influence of Life a tree is compounded of the elements of what is commonly called soda water, put together with the aid of sunshine; the solar energy is directed to that end and to the production of the oxygen essential for animal life on this planet. The energy is derived from the sun, but is controlled and directed by life into structures which otherwise would be impossible: an analogy is the directing power of an organist over the energy supplied by the bellows; he arranges it into music. So also a formless mass of wax becomes changed into a honeycomb, by the agency of life. The stones of a quarry are arranged and converted into a cathedral, and an element of beauty is added by the designing power of mind. Wherever we see order and beauty we may know that mind has been at work. A mindless operation, such as often occurs in the inorganic world, usually results in an increase of disorder and mere random confusion. So much so that some hold that disorganisation is bound to occur sooner or later, and that that must be the end of the present cosmos. But they are leaving out the controlling, guiding, ordering effect of mind. Clerk Maxwell showed how the activity of sheer intelligence could sort out confusion into order, could redistribute heat which had run down into inequality of temperature, and could undo the effects of the second law of thermodynamics. People are apt to make too much of that law; it only applies to inanimate physical agency. The physical universe left to itself may be running down; but mind can reorganise it, can reverse the process, or can start it afresh.

If we are raising stones to form part of a structure, it takes the same amount of work to place them in positions of ugliness as in places where they will add to the beauty of the whole. This is the result of design. Can we not see evidence of similar design in a bird's feather, an

THE BEARING ON RELIGION

insect's wing, aye, even in a crystal structure? The design is deep-seated, not obvious as it is in the work of a human artificer. The things as it were make themselves—measures are taken to that end—a still higher feat of architecture; but they positively shout that Mind has been ultimately responsible for their organisation.

Why do we not bring theology more prominently into science? For a very good reason. It would be shirking the issue, it would be jumping all the intermediate steps. Everything is done by God; but it is our privilege to find out how; to understand the mode of working. Mind does not act directly, it acts through certain processes and intermediate stages which can be understood. The mechanism seems to run of itself: that is what perfect mechanism often seems to do, to a superficial observer. The business of the scientific enquirer is to ferret out the details of the mechanism, whether it be chemical mechanism or any other, and to ascertain its object. Every result has a cause which can be traced. We can point out the stages of the process, we can trace the operation of the secretions which bring a result to pass. To decline to do this would be to throw up the sponge and admit defeat. Sooner or later we may have to admit that we can penetrate no further, but we postpone the collapse of our scrutiny as long as we can. We are conscious of some planning and designing power in ourselves, aye, and of some creative power. A poem or a drama or a work of art is in a sense a creation. It had not previously existed, its parts were put together and arranged in due order by a mental effort. We can learn from that, in infantile fashion, what creation feels like. The greater the Artist the more he is hidden, concealed in his work. Not much is known about Homer or Shakespeare. We can hardly follow all the steps by which they proceeded on their way to the small

307

kind of perfection which they managed to achieve. How can we hope to follow all the operations of the Creator of the Universe save in a spirit of awe and reverence, that is, in a spirit of religion? Science probes and investigates, religion accepts and worships. There is room for both, in different moods. If we attempt to mix them there is confusion, there may appear to be conflict. Some people avoid the conflict by keeping the two moods or atmospheres distinct. That is legitimate enough. But if we can contemplate the whole in a spirit of unification, we shall attain a calmer and nobler philosophic standard, more worthy of our human attributes, more akin, we may conjecture, to the Divine.

We have learnt even in physics that there are mysterious guiding entities. We call them waves, or we call them *psi*, and have begun to deal with them, though we do not know what they are. I am inclined to speculate and say that these things of which the first glimpse has been caught by recent physics may be part of the manifestation of life and mind, and that it is by their aid that mind operates and guides events in the physical universe. This speculation may be wrong, but whether wrong or not, we may be certain that spiritual entities exist, and have far more to do with our actions and our thoughts, our hopes and sublimer feelings, than we have yet been able to imagine.

The unseen universe is a great reality, that is the region to which we really belong, and to which we shall one day return. We are only associated with matter for a time; we can use it thankfully while we are here, but need not make the mistake of assuming that it is all that exists. In ourselves we know better. A church in every village testifies to belief in the existence of a spiritual world. We are still groping after God if haply we may find Him. Let us not be perturbed by the mechanistic teaching of science, but accept it for what

THE BEARING ON RELIGION

it is, a true and laborious attempt to interpret the meaning of the things around us, a finding of pebbles on the beach, as Newton said, while the whole ocean of truth extends unexplored before us.

So far I have dealt with religion only in general terms. But it seems to me that what I have said about the functions of matter are very applicable to such a problem as that of incarnation. We are all conscious of being spirits, we do not know in what form we previously existed, but we are sure that we shall continue to exist, and that meanwhile we utilise the matter of this planet for an episode of education and struggle and effort; having freedom to obey or disobey the laws which we find in operation here, and on the whole suffering if we disobey them. We have learnt something about the nature of this matter, which is to us a foreign body somewhat difficult to deal with. We do not know very much about it, though for some three centuries we have made heroic attempts to understand; we find that it is built of atoms, which are not indivisible specks as we used to think, but are built of electrical particles on the pattern of a solar system; also that they have the power under certain circumstances of converting themselves into a flash of radiation, and so becoming dematerialised. The animated particles which constitute our bodies can do many surprising things, can perform heroic acts, can display self-sacrifice and human feeling and love and many of our higher attributes. So much so that even our bodies by their structure display something of the purposes to which they have been put: so that a saint or a great man can become an object of veneration even in his bodily form. How far that can be carried we do not know. The soul constructs the body, and a mighty soul may have an influence over the body such as we ordinary folk can hardly imagine.

MY PHILOSOPHY.—PART IV

Our belief is that there was one Personality who chose to become incarnate in matter some nineteen hundred years ago, for the purpose partly perhaps of acquiring experience of that state of existence, but mainly for the sake of helping those who thus became his brethren; and lived such a life that the very matter of his body became on a certain occasion transfigured and shone with an unearthly light. We are also taught, and some of us believe, that when by the priests and orthodox people of his day he was put to death with the utmost ignominy, his body was so transfused with the spirit which had animated it that it dematerialised and left the tomb empty. There is nothing in that which seems to me impossible or incompatible with the line of future discovery about material processes and the influence of the spirit on the body. I have often wondered what instinct it was that caused the Church, and let us say that good and enlightened man Bishop Gore, to attach so much importance and to concentrate so essentially on the empty tomb. It does not help the doctrine of survival: surely in preaching about our survival as analogous to the Resurrection, the empty tomb constitutes a real difficulty. Our tombs will not be empty: though the doctrine led people to imagine that in some future day their tombs would be empty too, that their discarded bodies would be resuscitated and once more animated by the rejoining spirit. That is clearly false.

Why then should His tomb have been empty? Why did his resurrection differ from ours? Is it that he anticipated a future higher grade of mankind? Was his spirit so high that it not only animated the body, but changed it, altered the perceptible material form, so that in a literal sense he became the firstfruits of them that slept? And is it some germ of this perception that has led the Church to formulate the doctrine of a bodily resurrection, at least in his case? It has nothing to do

THE BEARING ON RELIGION

with the forty-day Appearances: the old tortured body was not necessary for them. But it seems to me quite possible that his case was an anticipation of what in time may happen to many, that after a long course of evolution our bodies too may become dematerialised, and that all the repulsive paraphernalia of burial or burning, to get rid of the unwholesome residue of de-organising or disintegrating matter that we leave behind, shall no longer be necessary. Not that our bodies will rejoin the spirit, the spirit will not need them, it will have a spiritual or etheric body of its own. Our present material bodies are formed of earthly particles, and to the earth they will always return; but perhaps they need not always go through the processes of decomposition which to many are so repulsive. The atoms themselves may separate and so spontaneously disappear from our ken; and the body, having served its purpose, may be not only discarded, but may cease to be. I do not know if this will ever be the fate of the higher portions of humanity, it is a long time ahead yet anyhow, but we need not shut our eyes to the possibility. And if we find the evidence good, we may adhere to our faith that our Elder Brother had already attained this high eminence, and that the tomb could not hold the body which had been animated by so lofty a spirit.

The human race has a long time ahead of it; astronomers tell us the planet will still be habitable millions of years hence. Evolution is still going on, and what changes may occur in all that time we little know. But I am now speculating beyond the bounds of physical science, I do not know what is possible and what is impossible. All I can say is that I see no reason to doubt the possibility, and that if our faith and intuition lead us in any such direction, we need not assume that our present knowledge of the universe is sufficient to enable us to deny it, or to pour scorn upon those who hold the belief. Let us not be dogmatic either way. Our Master undoubtedly pre-existed as the Eternal Christ, and is as living and active to-day as ever He was, having acquired the power of omnipresence and many other faculties of which we have no present knowledge. He lived on earth for a short time as Jesus of Nazareth, and met with that rejection and contumely which awaits all pioneers; but already he has influenced and redeemed the world to an amazing extent. All the meaning and consequences of that Incarnation we are not likely to know, from any arguments based on scientific procedure. We can be thankful that he has revealed to us part of the nature of the Deity whose power and majesty are revealed by science, but who has other attributes of love and simplicity and affection. These truly human attributes of God were revealed by Christ. He and the Father were one in plan and intention; he was perfectly obedient to his Father's will. He foresaw that only thus could the Kingdom of Heaven arrive upon earth. His prayer was, and it is ours too, Thy will be done, Thy Kingdom come.

THE END

INDEX

INDEX

INDEX

315

INDEX

316

INDEX

INDEX

MADE AND PRINTED IN GREAT BRITAIN BY PURNELL AND SONS
PAULTON (SOMERSET) AND LONDON